Methods in Computational Chemistry

Volume 4
Molecular Vibrations

METHODS IN COMPUTATIONAL CHEMISTRY

Volume 1 Electron Correlation in Atoms and Molecules
Edited by Stephen Wilson

Volume 2 Relativistic Effects in Atoms and Molecules
Edited by Stephen Wilson

Volume 3 Concurrent Computation in Chemical Calculations
Edited by Stephen Wilson

Volume 4 Molecular Vibrations
Edited by Stephen Wilson

A Continuation Order Plan is available for this series. A continuation order will bring delivery of each new volume immediately upon publication. Volumes are billed only upon actual shipment. For further information please contact the publisher.

Methods in Computational Chemistry

Volume 4
Molecular Vibrations

Edited by

STEPHEN WILSON
Rutherford Appleton Laboratory
Oxfordshire, England

Plenum Press • New York and London

Library of Congress Cataloging in Publication Data

(Revised for volume 4)
Methods in computational chemistry.
Computer disk (5¼ in.) in pocket of v. 2; requires IBM PC AT, 386, or compatible machine; 640K RAM; MS-DOS; double-sided, high density (1.2M) disk drive; 8087 arithmetic processor.
Includes bibliographies and indexes.
Contents: v. 1. Electron correlation in atoms and molecules / edited by Stephen Wilson—v. 2. Relativistic effects in atoms and molecules / edited by Stephen Wilson—[etc.]—v. 4. Molecular vibrations / edited by Stephen Wilson.
1. Chemistry—Data processing. I. Wilson, S. (Stephen), 1950–
QD39.3.E46M47 1987 542 87-7249
ISBN 0-306-44168-3

ISBN 0-306-44168-3

A Division of Plenum Publishing Corporation
233 Spring Street, New York, N.Y. 10013

Printed in the United States of America

Contributors

James R. Henderson, Department of Physics and Astronomy, University College London, London WC1E 6BT, U.K.

Ivan Hubač, Division of Chemical Physics, Faculty of Mathematics and Physics, Komensky University, 842 15 Bratislava, Czechoslovakia

Steven Miller, Department of Physics and Astronomy, University College London, London WC1E 6BT, U.K.

Brian T. Sutcliffe, Department of Chemistry, University of York, York YO1 5DD, U.K.

Michal Svrček, Institute of Chemistry, Division of Chemistry and Physics of Biomolecules, Komensky University, 832 33 Bratislava, Czechoslovakia

Jonathan Tennyson, Department of Physics and Astronomy, University College London, London WC1E 6BT, U.K.

Stephen Wilson, Rutherford Appleton Laboratory, Chilton, Oxfordshire OX11 0QZ, U.K.

From the Preface to Volume 1

Today the digital computer is a major tool of research in chemistry and the chemical sciences. However, although computers have been employed in chemical research since their very inception, it is only in the past ten or fifteen years that computational chemistry has emerged as a field of research in its own right. The computer has become an increasingly valuable source of chemical information, one that can complement and sometimes replace more traditional laboratory experiments. The computational approach to chemical problems can not only provide a route to information that is not available from laboratory experiments but can also afford additional insight into the problem being studied, and, as it is often more efficient than the alternative, the computational approach can be justified in terms of economics.

The applications of computers in chemistry are manifold. A broad overview of both the methods of computational chemistry and their applications in both the industrial research laboratory and the academic research environment is given in my book *Chemistry by Computer* (Plenum Press, 1986). Applications of the techniques of computational chemistry transcend the traditional divisions of chemistry—physical, inorganic, and organic—and include many neighboring areas in physics, biochemistry, and biology. Numerous applications have been reported in fields as diverse as solid-state physics and pesticide research, catalysis and pharmaceuticals, nuclear physics and forestry, interstellar chemistry and molecular biology, and surface physics and molecular electronics. The range of applications continues to increase as research workers in chemistry and allied fields identify problems to which the methods of computational chemistry can be applied.

The techniques employed by the computational chemist depend on the size of the system being investigated, the property or range of properties of interest, and the accuracy to which these properties must be measured. The methods of computational chemistry range from quantum-mechanical studies of the electronic structure of small molecules to the determination of bulk properties by means of Monte Carlo or molecular dynamics simulations; from the study of protein structures using the methods of molecular mechanics to the investigation of simple molecular collisions; from expert systems for the design of synthetic routes in organic chemistry to the use of computer graphics techniques to investigate interactions between biological molecules.

The computers employed in chemical calculations vary enormously, from the small microcomputers used for data analysis to the large state-of-the-art machines that are frequently necessary for contemporary ab initio calculations of molecular electronic structure. Increasingly, large mainframe computers are departing from the traditional von Neumann architecture with its emphasis on serial computation, and a similar change is already underway in smaller machines. With the advent of vector processing and parallel processing computers, the need to match an algorithm closely to the target machine has been recognized. Whereas different implementations of a given algorithm on traditional serial computers may lead to programs that differ in speed by a factor of about 2, factors of 20 were not uncommon with the first vector processors, and larger factors can be expected in the future.

With the increasing use of computational techniques in chemistry, there is an obvious need to provide specialist reviews of methods and algorithms so as to enable the effective exploitation of the computing power available. This is the aim of the present series of volumes. Each volume will cover a particular area of research in computational chemistry and will provide a broad-ranging yet detailed analysis of contemporary theories, algorithms, and computational techniques. The series will be of interest to those whose research is concerned with the development of computational methods in chemistry. More importantly, it will provide an up-to-date summary of computational techniques for the chemist, atomic and molecular physicist, biochemist, and molecular biologist who wish to employ the methods to further their research programs. The series will also provide the graduate student with an easily accessible introduction to the field.

Preface

Physically, it is expected that the large disparity between the mass of the electrons and the mass of the nuclei in a molecule will allow the electronic motion to accommodate almost instantaneously any change in the positions of the nuclei. Consequently, the electronic and nuclear motion can be treated separately to a very good approximation. The electrons experience the nuclei as fixed force centers and adiabatically follow any change in the nuclear positions. Conversely, the nuclei experience an averaged interaction with the electrons. This is the essence of the Born–Oppenheimer approximation. The separation of the nuclear and electronic motions is almost invariably the first set in any application of quantum mechanics to molecules. It allows the total energy of a molecular system to be written as a sum of the energy associated with electronic motion and the energy associated with the nuclear motion.

This volume is concerned with the description of the nuclear motion in isolated molecules. Such motion is largely associated with the vibrations of the nuclei about their equilibrium positions together with the rotation of the molecule as a whole. In the first chapter of the present volume, I provide an introduction to molecular vibration theory which should provide the necessary background for the following articles. The mathematics of vibration–rotation calculations is considered in detail by Brian T. Sutcliffe in the second chapter. In Chapter 3, Jonathan Tennyson, Steven Miller, and James R. Henderson describe the calculation of highly excited vibration-rotation states of triatomic molecules. Finally, in Chapter 4, the many-body perturbation theory of the vibrational–electronic problem in molecules is described by Ivan Hubač and Michal Svrček.

The study of molecular vibrations provides an important bridge between theoretical studies of molecular structure and experimentally observed vibration and vibration–rotation spectra. Together, the four chapters in this volume provide a broad-ranging yet thorough analysis of the most important aspects of contemporary research into the molecular vibration problem.

Stephen Wilson

Wood End

Contents

1. An Introduction to Molecular Vibration Theory

Stephen Wilson

2. The Mathematics of Vibration–Rotation Calculations

Brian T. Sutcliffe

3. The Calculation of Highly Excited Vibration-Rotation States of Triatomic Molecules

Jonathan Tennyson, Steven Miller, and James R. Henderson

4. The Many-Body Perturbation Theory of the Vibrational–Electronic Problem in Molecules

Ivan Hubač and Michal Svrček

1

An Introduction to Molecular Vibration Theory

STEPHEN WILSON

1. Introduction

The large disparity between the mass of the electrons and the mass of the nuclei in a molecule allows the electronic motion to accommodate almost instantaneously any change in the positions of the nuclei. Consequently, both electronic and nuclear motion can be treated separately to a very good approximation, and the total energy may be written as the sum

$$E_{\text{total}} = E_{\text{electron}} + E_{\text{nuclear}} \tag{1}$$

where E_{electron} and E_{nuclear} are the energies associated with electronic and nuclear motion, respectively. The corresponding wave function takes the form

$$\Psi_{\text{total}} = \Phi_{\text{electron}}\chi_{\text{nuclear}} \tag{2}$$

The electrons experience the nuclei as fixed-force centers and adiabatically follow any change in the nuclear positions. Conversely, the nuclei experience

STEPHEN WILSON • Rutherford Appleton Laboratory, Chilton, Oxfordshire OX11 0QX, U.K.

Methods in Computational Chemistry, Volume 4: Molecular Vibrations, edited by Stephen Wilson, Plenum Press, New York, 1992.

an averaged interaction with the electrons. This is the essence of the Born–Oppenheimer approximation. The separation of nuclear and electronic motion is almost invariably the first step in any application of quantum mechanics to molecules.

This volume describes nuclear motion in isolated molecules. Nuclear motion is largely associated with the vibrational motions of the nuclei about their equilibrium positions, together with the rotation of the molecule as a whole. This chapter presents an introduction to molecular vibration theory with an overview of the "traditional" approach to the molecular vibration problem in molecules. The aim here is to provide some background material for the more advanced presentations given in the following chapters. More detailed expositions may be found in the standard texts such as Wilson *et al.*[1] and Herzberg.[2]

The present chapter is structured as follows: Section 2 considers vibrational motion in diatomic molecules and the interaction between vibrational and rotational motion for these systems. Section 3 considers polyatomic molecules, especially the basic molecular model employed, coordinate systems, the Eckart conditions, kinetic energy, and the potential energy function. The secular equation arising in the molecular vibration problem is described in Section 4 where mass-adjusted coordinates are introduced in Section 4.1 and normal coordinates in Section 4.2. Section 5 examines internal coordinates, discussing Wilson's S vectors, the exploitation of point group symmetry, and a simple example (the H_2O molecule). Finally, Section 6 contains concluding remarks.

2. Vibrations in Diatomic Molecules

Diatomic molecules provide a simple introduction to the theory of vibrations and vibration–rotation interaction in molecules. Solution of the electronic Schrödinger equation yields the electronic energy of a molecule for the particular electronic state under consideration for a given position of the nuclei. The electronic energy thus depends parametrically on the nuclear coordinates. These electronic energy values, when regarded as a function of the nuclear configuration, give the potential energy curve or surface on which the nuclei move.

For a diatomic molecule, the potential energy curve may be written

$$U = U(R) \tag{3}$$

where R is the internuclear separation. This function is included in the

Schrödinger equation for the nuclear motion, which takes the form

$$[-(2\mu)^{-1}\nabla^2 + U(R)]\Psi_{\text{nuclear}} = E\Psi_{\text{nuclear}} \tag{4}$$

μ is the reduced mass, which is defined as

$$\mu = M_A M_B (M_A + M_B)^{-1} \tag{5}$$

where M_A is the mass of nucleus A. The nuclear Schrödinger equation may be separated by using spherical polar coordinates and putting

$$\Psi_{\text{nuclear}} = R^{-1}\chi(R)\mathscr{Y}_{l,m}(\theta, \phi) \tag{6}$$

where $\mathscr{Y}_{l,m}(\theta, \phi)$ is a spherical harmonic. This substitution leads to a radial equation of the form

$$\{-(2\mu)^{-1}\nabla^2 + U(R) + (2\mu R^2)^{-1}[J(J+1)]\}\chi(R) = E\chi(R) \tag{7}$$

with $J = 0, 1, 2, \ldots$ being the rotational quantum number. The nuclear wave function is thus separated into a vibrational function and a rotational function.

If the potential energy curve for a diatomic molecule has been calculated, the vibrational and rotational levels may then be obtained, in principle, by solution of the nuclear wave equation. Terms neglected in the separation of the electronic and nuclear parts of the problem give rise to small diagonal corrections which may be added to the potential energy function. The nuclear Schrödinger equation for diatomic molecules can be solved directly by numerical integration, following the approach first described by Cooley[3] and Cashion.[4] The drawback of this approach for obtaining the nuclear wave equation is that a very large number of points on the potential energy curve need to be used in order to obtain the required accuracy.

A simpler approach is to solve the vibrational equation

$$\left[(2\mu)^{-1}\frac{\partial^2}{\partial R^2} + U(R)\right]\chi(R) = E\chi(R) \tag{8}$$

and treat the term $(2R^2)^{-1}[J(J+1)]$ as a small perturbation. The vibrational equation can be solved either by the Cooley–Cashion method or by expanding $\chi(R)$ in some set of functions and invoking the variation theorem.

Usually, the approximate solutions of the electronic Schrödinger equation are obtained in tabular form at a relatively small number of internuclear distances. In order to consider vibrational and rotational effects, it is necessary to interpolate between the calculated values. Power series expansions afford a reasonably general means of doing this.

There are a number of factors that can influence the expansion coefficients and hence the calculated vibrational frequencies, rotation constants, and equilibrium geometries in a power series expansion for the potential energy function. These factors include:

(a) the accuracy of the calculated points;
(b) the order of the polynomial fitted;
(c) the range, number, and distribution of the calculated points;
(d) the availability of first- and higher-order derivatives; and
(e) the expansion variable involved.

For the first two factors, note that there is some advantage to fitting different orders of polynomial both to energies obtained by means of an independent electron model, such as the matrix Hartree–Fock method, and also to the correlation energy corrections. Calculated correlation energies may be more susceptible to error than the Hartree–Fock reference energies. The energy expression for the independent electron model contains a relatively small number of terms, whereas the correlation energy involves contributions from all parts of the reference spectrum. It is sometimes useful therefore, to interpolate the calculated correlation energies by a lower-order polynomial than is used for the reference energies. Of course, the nuclear repulsion energy is known for all configurations of the nuclei. There is often some advantage in fitting a polynomial to the calculated electronic energy values and then adding to this the analytic expression for the nuclear repulsion effects.

Factors (c) and (d) are closely related. If first- and higher-order derivatives of the energy with respect to nuclear positions are available, then a higher accuracy can be achieved with a smaller number of points than would be required using energy values only. As computers become increasingly powerful and algorithms more sophisticated, the limitation of the number of points available is being gradually eliminated. Sometimes, however, it is useful to carry out a "sensitivity analysis," that is, to repeat the fitting procedure after omitting some point(s) and then observing the effects on the expansion coefficients.

There appear to be three useful choices of expansion variable [factor (e)] which can be usefully employed in the power series expansion for the potential energy function

$$U(\rho) = A_0\left(1 + \sum_{i=1} a_i\rho^i\right) \tag{9}$$

The coefficients A_0 and a_i are to be determined by a least-squares fitting procedure. The Dunham parameter[5] is

$$\rho_1 = \frac{r - r_e}{r_e} \tag{10}$$

where r is the internuclear separation, and r_e is its equilibrium value. By employing this parameter a power series is obtained which has a radius of convergence

$$0 < r < 2r_e \tag{11}$$

The modified Dunham parameter originally introduced by Fougere and Nesbet[6] but often attributed to Simons, Parr, and Finlan,[7]

$$\rho_2 = \frac{r - r_e}{r} \tag{12}$$

has a radius of convergence

$$\tfrac{1}{2}r_e < r < \infty \tag{13}$$

This parameter has been widely employed. On the other hand, the modified Dunham parameter[8,9]

$$\rho_3 = \frac{r - r_e}{r + r_e} \tag{14}$$

has received relatively little attention despite having a larger radius of convergence than either ρ_1 or ρ_2. For ρ_3, the radius of convergence is

$$0 < r < \infty \tag{15}$$

If the power series expansions corresponding to ρ_1, ρ_2, and ρ_3 are written in the forms

$$U(\rho_1) = A_0\left(1 + \sum_{i=1} a_i \rho_1^i\right) \tag{16}$$

$$U(\rho_2) = B_0\left(1 + \sum_{i=1} b_i \rho_2^i\right) \tag{17}$$

$$U(\rho_3) = C_0\left(1 + \sum_{i=1} c_i \rho_3^i\right) \tag{18}$$

respectively, then the coefficients in these expansions may be related by the equations[10]

$$A_0 = B_0 \tag{19a}$$

$$a_n = b_n + \sum_{i=1}^{n-1} (-1)^i \binom{n+1}{i} b_{n-i} + (-1)^n(n+1) \tag{19b}$$

and

$$A_0 = \tfrac{1}{4} C_0 \tag{20a}$$

$$a_n = \left(\frac{1}{2}\right)^n \left[c_n + \sum_{i=1}^{n-1} (-1)^i \binom{n+1}{i} c_{n-i} + (-1)^n(n+1)\right] \tag{20b}$$

By using the expansion parameters ρ_1, ρ_2, and ρ_3, it has been assumed that r_e is known. In general, however, this is not the case; therefore, an iterative procedure has to be adopted starting from some initial guess for r_e. The coefficients a_i, b_i, and c_i are just the derivatives of the potential energy function:

$$a_i = (i!)^{-1}\left(\frac{\partial^i U(\rho_1)}{\partial \rho_1^i}\right)_{\rho_1 = 0} \tag{21}$$

$$b_i = (i!)^{-1}\left(\frac{\partial^i U(\rho_2)}{\partial \rho_2^i}\right)_{\rho_2 = 0} \tag{22}$$

$$c_i = (i!)^{-1}\left(\frac{\partial^i U(\rho_3)}{\partial \rho_3^i}\right)_{\rho_3 = 0} \tag{23}$$

It is obvious that $a_1 = b_1 = c_1 = 0$, and so the leading term, beyond the constants A_0, B_0, and C_0, is the quadratic

$$a_2\rho_1^2, \quad b_2\rho_2^2, \quad \text{or} \quad c_2\rho_3^2 \tag{24}$$

The leading term of the Hamiltonian is therefore a one-dimensional harmonic oscillator whose eigenvalues and eigenfunctions can be written down by inspection. The effects of the higher-order terms in the potential energy function on the vibration problem may determined by perturbation theory. We have already noted that the rotational terms can also be treated by perturbative methods. The result of this analysis is a series of formulae which relate the coefficients a_i, b_i, and c_i to the coefficients in the empirical representation of the spectrum by a convergent power series expansion in both the vibrational quantum number, v, and the rotational quantum number, J.

$$\begin{aligned}
\frac{E_{v,J}}{hc} &= G(v) + F_v(J) \\
G(v) &= \omega_e(v + \tfrac{1}{2}) - \omega_e x_e(v + \tfrac{1}{2})^2 + \omega_e y_e(v + \tfrac{1}{2})^3 + \cdots \\
F_v(J) &= B_v J(J+1) - D_v J^2(J+1)^2 + H_v J^3(J+1)^3 + \cdots \\
B_v &= B_e - \alpha^B(v + \tfrac{1}{2}) + \gamma^B(v + \tfrac{1}{2})^2 + \cdots \\
D_v &= D_e - \beta(v + \tfrac{1}{2}) + \cdots
\end{aligned} \tag{25}$$

In 1932, Dunham[5] calculated the energy levels of a rotating vibrator using the Wentzel–Kramers–Brillouin method. Using the expansion parameter, ρ_1, Dunham obtained the energy level formulae

$$E_{v,J} = \sum_{\ell,j} Y_{\ell,J}(v + \tfrac{1}{2})^\ell [J(J+1)]^j \tag{26}$$

where the first subscript under Y refers to the power of the vibrational quantum number and the second that of the rotational quantum number. The Y's are commonly referred to as the Dunham coefficients. The first 15

Dunham coefficients are displayed in Table 1. The Dunham coefficients are not exactly equal to the spectral constants obtained by means of perturbation theory. The connection between the Dunham coefficients and the band spectrum constants is as follows:

$$\begin{array}{lll} Y_{10} \simeq \omega_e & Y_{20} \simeq -\omega_e x & Y_{30} \simeq -\omega_e y \\ Y_{01} \simeq B_e & Y_{11} \simeq -\alpha_e & Y_{21} \simeq \gamma_e \\ Y_{02} \simeq D_e & Y_{12} \simeq B_e & Y_{40} \simeq \omega_e z \\ Y_{03} \simeq F_e & Y_{13} \simeq H_e & \end{array} \tag{27}$$

3. Polyatomic Molecules

3.1. The Molecular Model

We turn now to the much more complicated problem of describing vibration and vibration–rotation interaction in nonlinear polyatomic molecules. A molecular model is employed consisting of point masses, the atomic nuclei, bound together by the surrounding electrons and associated electrostatic forces. We shall assume that the nuclei are always close to their equilibrium positions in the molecule. The electrons and electrostatic forces only concern us to the extent that they define a potential energy function, U, which is a unique function of the internal molecular configuration, with a fairly deep minimum compared with kT, at the equilibrium configuration. This is essentially the Born–Oppenheimer approximation.[11,12]

This molecular model has degrees of freedom corresponding to overall translation, overall rotation, and internal vibration. We will study the nuclear motion problem using quantum mechanics. However, it is useful to first consider the problem classically and then to translate it into quantum-mechanical formalism, because a lot of the results are very similar and the translation is quite straightforward. The classical model affords a "picture" of the nuclear motion.

For realistic molecular models, overall translation, overall rotation, and internal vibration are almost independent of each other. In fact, the translation is completely independent of the other nuclear motion. The energy associated with nuclear motion may thus be written

$$E_{\text{nuclear}} = E_{\text{translation}} + E_{\text{vibration}} + E_{\text{rotation}} \tag{28}$$

Table 1. The Dunham Coefficients

$$Y_{00} = (B_e/8)(3a_2 - 7a_1^2/4)$$

$$Y_{10} = \omega_e\{1 + (B_e^2/4\omega_e^2)(25a_4 - 95a_1a_3/2 - 67a_2^2/4 + 459a_1^2a_2/8 - 1155a_1^4/64)\}$$

$$Y_{20} = (B_e/2)\{3(a_2 - 5a_1^2/4) + (B_e^2/2\omega_e^2)(245a_6 - 1365a_1a_5/2 - 885a_2a_4/2 - 1085a_3^2/4 + 8535a_1^2a_4/8 + 1707a_2^3/8 + 7335a_1a_2a_3/4 - 23{,}865a_1^3a_3/16 - 62{,}013a_1^2a_2^2/32 + 239{,}985a_1^4a_2/128 - 209{,}055a_1^6/512)\}$$

$$Y_{30} = (B_e^2/2\omega_e)(10a_4 - 35a_1a_3 - 17a_2^2/2 + 225a_1^2a_2/4 - 705a_1^4/32)$$

$$Y_{40} = (5B_e^3/\omega_e^2)(7a_5/2 - 63a_1a_5/4 - 33a_2a_4/4 - 63a_3^2/8 + 543a_1^2a_4/16 + 75a_2^3/16 + 483a_1a_2a_3/8 - 1953a_1^3a_3/32 - 4989a_1^2a_2^2/64 + 23{,}265a_1^4a_2/256 - 23{,}151a_1^6/1025)$$

$$Y_{01} = B_e\{1 + (B_e^2/2\omega_e^2)(15 + 14a_1 - 9a_2 + 15a_3 - 23a_1a_2 + 21(a_1^2 + a_1^3)/2)\}$$

$$Y_{11} = (B_e^2/\omega_e)[6(1 + a_1) + (B_e^2/\omega_e^2)\{175 + 285a_1 - 335a_2/2 + 190a_3 - 225a_4/2 + 175a_5 + 2295a_1^2/8 - 459a_1a_2 + 1425a_1a_3/4 - 795a_1a_4/2 + 1005a_2^2/8 - 715a_2a_3/2 + 1155a_1^3/4 - 9639a_1^2a_2/16 + 5145a_1^2a_3/8 + 4677a_1a_2^2/8 - 14{,}259a_1^3a_2/16 - 31{,}185(a_1^4 + a_1^5)/128\}]$$

$$Y_{21} = (6B_e^3/\omega_e^2)\{5 + 10a_1 - 3a_2 + 5a_3 - 13a_1a_2 + 15(a_1^2 + a_1^3)/2\}$$

$$Y_{31} = (20B_e^4/\omega_e^3)\{7 + 21a_1 - 17a_2/2 + 14a_3 - 9a_4/2 + 7a_5 + 225a_1^2/8 - 45a_1a_2 + 105a_1a_3/4 - 51a_1a_4/2 + 51a_2^2/8 - 45a_2a_3/2 + 141a_1^3/4 - 945a_1^2a_2/16 + 435a_1^2a_3/8 + 411a_1a_2^2/8 - 1509a_1^3a_2/16 + 3807(a_1^4 + a_1^5)/128\}$$

$$Y_{02} = -(4B_e^3/\omega_e^2)[1 + (B_e^2/2\omega_e^2)\{163 + 199a_1 - 119a_2 + 90a_3 - 45a_4 - 207a_1a_2 + 205a_1a_3/2 - 333a_1^2a_2/2 + 693a_1^2/4 + 46a_2^2 + 126(a_1^3 + a_1^4/2)\}]$$

$$Y_{12} = -(12B_e^4/\omega_e^3)(19/2 + 9a_1 + 9a_1^2/2 - 4a_2)$$

$$Y_{22} = -(24B_e^5/\omega_e^4)\{65 + 125a_1 - 61a_2 + 30a_3 - 15a_4 + 495a_1^2/4 - 117a_1a_2 + 26a_2^2 + 95a_1a_3/2 - 207a_1^2a_2/2 + 90(a_1^3 + a_1^4/2)\}$$

$$Y_{03} = 16B_e^5(3 + a_1)/\omega_e^4$$

$$Y_{13} = (12B_e^6/\omega_e^5)(233 + 279a_1 + 189a_1^2 + 63a_1^3 - 88a_1a_2 - 120a_2 + 80a_3/3)$$

$$Y_{04} = -(64B_e^7/\omega_e^6)(13 + 9a_1 - a_2 + 9a_1^2/4)$$

There are small interactions, however, between the vibrational and rotational motion due to

(a) the centrifugal stretching of the molecule as it rotates, and
(b) the effect of Coriolis forces in polyatomic molecules.

These interactions are small and can usually be neglected to the first order both because the amplitudes of the internal vibrations are assumed to be small compared with the internuclear distances and because the frequencies of vibration are always large compared with the frequencies of molecular rotation.

3.2. Coordinate Systems

It is convenient to consider the coordinates of the atoms in a given molecular system with respect to two systems of axes—one fixed in space and the other fixed on the molecule, that is, translating and rotating with the molecule. The manner in which we fix the moving system is considered in some detail below. In fact, it is fairly obvious that the origin of the molecule fixed system must always be chosen to coincide with the center of mass of the molecule. It is not so obvious, however, as to how one should relate the rotation of the moving system to the rotating molecule.

Let us assume that the molecule contains N atoms labeled $p = 1, 2, \ldots, N$. Let the coordinates of atom p in the molecule fixed system be

$$(x_p, y_p, z_p) = \mathbf{r}_p \tag{29}$$

and let the equilibrium coordinates of the atom p be

$$(a_p, b_p, c_p) = \mathbf{a}_p \tag{30}$$

These coordinates are illustrated in Fig. 1, where we also show the displacement of atom p from the equilibrium position

$$(\Delta x_p, \Delta y_p, \Delta z_p) = \boldsymbol{\rho}_p \tag{31}$$

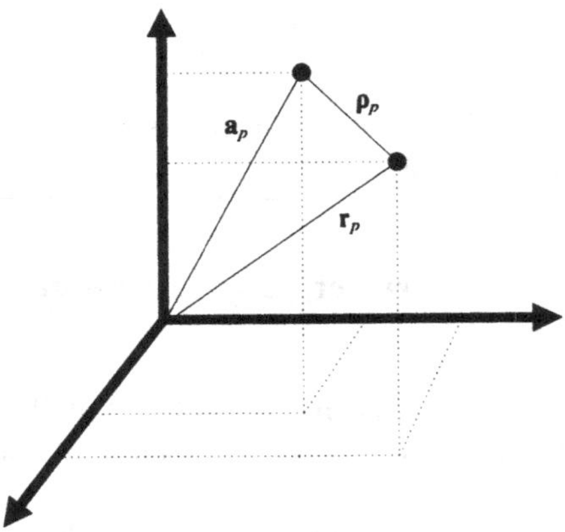

Figure 1. The molecule fixed coordinate system.

It is clear from Fig. 1 that

$$\boldsymbol{\rho}_p = \mathbf{r}_p - \mathbf{a}_p \tag{32}$$

The velocity of the atom p defined with respect to the moving coordinate system is given by

$$(\Delta\dot{x}_p, \Delta\dot{y}_p, \Delta\dot{z}_p) = (\dot{x}_p, \dot{y}_p, \dot{z}_p) = \left(\frac{dx_p}{dt}, \frac{dy_p}{dt}, \frac{dz_p}{dt}\right) = \mathbf{v}_p \tag{33}$$

Now, using Fig. 2, let us consider the relation between the molecule fixed coordinate system and the space fixed system. Let the origin of the moving system with respect to the space fixed system be

$$(X, Y, Z) = \mathbf{R} \tag{34}$$

and let

$$(\dot{\theta}_X, \dot{\theta}_Y, \dot{\theta}_Z) = \boldsymbol{\omega} \tag{35}$$

$$\dot{\theta}_X = \frac{d\theta_X}{dt}, \qquad \dot{\theta}_Y = \frac{d\theta_Y}{dt}, \qquad \dot{\theta}_Z = \frac{d\theta_Z}{dt} \tag{36}$$

be the angular velocity of the moving axes relative to the space fixed system. The position of atom p with respect to the space fixed system is

$$\boldsymbol{\Gamma}_p = \mathbf{R} + \mathbf{r}_p \tag{37}$$

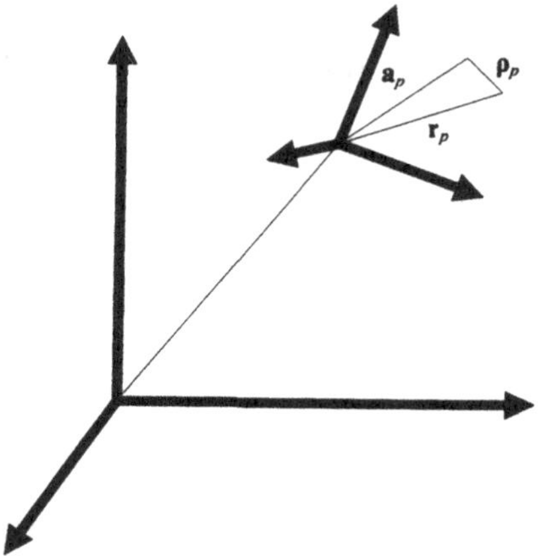

Figure 2. The relation between the molecule fixed coordinate system and the space fixed system.

The velocity of atom p relative to the space fixed coordinate system is then

$$\begin{aligned}\mathbf{u}_p = \dot{\mathbf{\Gamma}} &= \dot{\mathbf{R}} + \dot{\mathbf{r}}_p \\ &= \dot{\mathbf{R}} + \boldsymbol{\omega} \times \mathbf{r}_p + \mathbf{v}_p \end{aligned} \tag{38}$$

because

$$\dot{\mathbf{r}}_p = \boldsymbol{\omega} \times \mathbf{r}_p + \mathbf{v}_p \tag{39}$$

The total kinetic energy, T, which may be written

$$2T = \sum_p m_p \dot{u}_p^2 \tag{40}$$

is then given by

$$\begin{aligned} 2T = M\dot{\mathbf{R}}^2 &+ && (41a)\\ &\sum_p m_p(\boldsymbol{\omega} \times \mathbf{r}_p) \cdot (\boldsymbol{\omega} \times \mathbf{r}_p) + && (41b)\\ &\sum_p m_p v_p^2 + && (41c)\\ &2\dot{\mathbf{R}} \cdot \sum_p m_p \mathbf{v}_p + && (41d)\\ &2(\dot{\mathbf{R}} \times \boldsymbol{\omega}) \cdot \sum_p m_p \mathbf{r}_p + && (41e)\\ &2\boldsymbol{\omega} \cdot \sum_p m_p(\mathbf{r}_p \times \mathbf{v}_p) && (41f)\end{aligned}$$

where m_p is the mass of atom p and

$$M = \sum_p m_p \tag{42}$$

is the total mass of the nuclei. In expression (41), the identity

$$(a \times b) \cdot c = a \cdot (b \times c) \tag{43}$$

has been used to put

$$\dot{\mathbf{R}} \cdot (\boldsymbol{\omega} \times \mathbf{r}_p) = (\dot{\mathbf{R}} \times \boldsymbol{\omega}) \cdot \mathbf{r}_p \tag{44}$$

and

$$(\boldsymbol{\omega} \times \mathbf{r}_p) \cdot \mathbf{v}_p = \boldsymbol{\omega} \cdot (\mathbf{r}_p \times \mathbf{v}_p) \tag{45}$$

3.3. The Eckart Conditions

We have not yet completely defined the moving coordinate system. Although it only requires $3N$ coordinates to define the configuration of the molecular model in space, we are employing six coordinates, three translational and three rotational, to fix the moving axis system relative to the space fixed coordinates, and a further $3N$ coordinates to define the positions of the atoms relative to the moving system. Thus, we have a total of $3N + 6$ coordinates all of which cannot be independent. We must impose six conditions,[13,14] relating the coordinates (x_p, y_p, z_p), which will serve to completely define the moving system of axes in relation to the molecule.

If we put

$$\sum_p m_p \mathbf{r}_p = 0 \tag{46}$$

that is,

$$\sum_p m_p x_p = 0, \qquad \sum_p m_p y_p = 0, \qquad \sum_p m_p z_p = 0 \tag{47}$$

then three conditions are fixed (one for each coordinate). This condition states that the origin of the moving coordinate system should coincide with the center of mass of the molecular model. It also implies

$$\sum_p m_p \dot{\mathbf{r}}_p = 0 \tag{48}$$

and because

$$\dot{\mathbf{r}}_p = \boldsymbol{\omega} \times \mathbf{r}_p + \mathbf{v}_p \tag{49}$$

we have

$$\sum_p m_p(\boldsymbol{\omega} \times \mathbf{r}_p) + \sum_p m_p \mathbf{v}_p = 0 \tag{50}$$

and thus

$$\sum_p m_p \mathbf{v}_p = 0 \tag{51}$$

It follows that the condition (46) makes the fourth and fifth terms on the right-hand side of (41) equal to zero.

The remaining three Eckart conditions are

$$\sum_p m_p(\mathbf{a}_p \times \mathbf{v}_p) = 0 \tag{52}$$

that is,

$$\sum_p m_p(b_p \dot{z}_p - c_p \dot{y}_p) = 0 \tag{53a}$$

$$\sum_p m_p(c_p \dot{x}_p - a_p \dot{z}_p) = 0 \tag{53b}$$

$$\sum_p m_p(a_p \dot{y}_p - b_p \dot{x}_p) = 0 \tag{53c}$$

These conditions do not make the sixth term of (41) zero, but they do make it small to first order. Because $\mathbf{r}_p = \mathbf{a}_p + \boldsymbol{\rho}_p$, equations (53) imply that

$$\sum_p m_p(\mathbf{r}_p \times \mathbf{v}_p) = \sum_p m_p(\boldsymbol{\rho}_p \times \mathbf{v}_p) \tag{54}$$

and we shall assume that displacements from equilibrium, $\boldsymbol{\rho}_p$, are always small.

3.4. The Kinetic Energy

Introducing conditions (46) and (52) into expression (41) for the kinetic energy, we obtain

$$2T = M\dot{R}^2 + \tag{55a}$$

$$\sum_p m_p(\boldsymbol{\omega} \times \mathbf{r}_p) \cdot (\boldsymbol{\omega} \times \mathbf{r}_p) + \tag{55b}$$

$$\sum_p m_p v_p^2 + \tag{55c}$$

$$2\boldsymbol{\omega} \cdot \sum_p m_p(\boldsymbol{\rho}_p \times \mathbf{v}_p) \tag{55d}$$

The first term in (55) is the overall kinetic energy of translation

$$2T_{\text{trans}} = M\dot{R}^2 \tag{56}$$

The second term is the overall kinetic energy of rotation. It may be written in the form

$$\begin{aligned} 2T_{\text{rot}} &= \sum_p m_p(\boldsymbol{\omega} \times \mathbf{r}_p)^2 \\ &= \sum_p m_p[(\omega_y z_p - \omega_z y_p)^2 + (\omega_z x_p - \omega_x z_p)^2 + (\omega_x y_p - \omega_y x_p)^2] \end{aligned} \tag{57}$$

which may be rearranged to give

$$\begin{aligned} 2T_{\text{rot}} = {} & \omega_x^2 \sum_p m_p(y_p^2 + z_p^2) + \omega_y^2 \sum_p m_p(x_p^2 + z_p^2) + \omega_z^2 \sum_p m_p(x_p^2 + y_p^2) \\ & - 2\omega_y \omega_z \sum_p m_p y_p z_p - 2\omega_z \omega_x \sum_p m_p z_p x_p - 2\omega_x \omega_y \sum_p m_p x_p y_p \end{aligned} \tag{58}$$

or

$$\begin{aligned} 2T_{\text{rot}} &= A\omega_x^2 + B\omega_y^2 + C\omega_z^2 - 2D\omega_y\omega_z - 2E\omega_z\omega_x - 2F\omega_x\omega_y \\ &= (\omega_x \quad \omega_y \quad \omega_z)\begin{pmatrix} A & -F & -E \\ -F & B & -D \\ -E & -D & C \end{pmatrix}\begin{pmatrix} \omega_x \\ \omega_y \\ \omega_z \end{pmatrix} \end{aligned} \tag{59}$$

where A, B, C and D, E, F are the moments and products of inertia. It should be emphasized that they are the instantaneous moments and products of inertia; that is, they are functions of the instantaneous distorted configuration. However, if we neglect displacements from equilibrium, then to a good approximation we can replace the instantaneous moments and products of inertia, $A, B, \ldots$, etc., by the moments and products of inertia for the equilibrium configuration

$$A_0 = \sum_p m_p(b_p^2 + c_p^2) \tag{60a}$$

$$B_0 = \sum_p m_p(a_p^2 + c_p^2) \tag{60b}$$

$$C_0 = \sum_p m_p(a_p^2 + b_p^2) \tag{60c}$$

$$D_0 = \sum_p m_p b_p c_p \tag{60d}$$

$$E_0 = \sum_p m_p a_p c_p \tag{60e}$$

$$F_0 = \sum_p m_p a_p b_p \tag{60f}$$

Furthermore, by choosing the axes to coincide with the principal axes of inertia in the equilibrium configuration

$$A_0 = I_x^0, \qquad B_0 = I_y^0, \qquad C_0 = I_z^0 \tag{61}$$

and

$$D_0 = E_0 = F_0 = 0 \tag{62}$$

the kinetic energy of rotation may be written in the form

$$2T_{\text{rot}} = I_x^0 \omega_x^2 + I_y^0 \omega_y^2 + I_z^0 \omega_z^2 \tag{63}$$

The third term in (55) is the kinetic energy of the internal vibration:

$$2T_{\text{vib}} = \sum_p m_p v_p^2 \tag{64}$$

Note that the $3N$ vectors, $\mathbf{v}_p$, are not all independent. By introducing the Eckart conditions, (46) and (52), the number of independent internal coordinates is reduced to $3N - 6$ for a polyatomic molecular model ($3N - 5$ for a linear species).

The fourth term in (55) arises from the coriolis interaction between the rotation and vibration

$$2T_{\text{vib-rot}} = 2\boldsymbol{\omega} \cdot \sum_p m_p(\boldsymbol{\rho}_p \times \mathbf{v}_p) \tag{65}$$

It is small to first order if $\boldsymbol{\rho}_p$ is small and may be neglected to the same approximation that the kinetic energy of rotation may be written in the form (63).

Thus, a good approximation to the kinetic energy for the nuclear motion is often

$$2T_{\text{total}} = M\dot{R}^2 + I_x^0\omega_x^2 + I_y^0\omega_y^2 + I_z^0\omega_z + \sum_p m_p v_p^2 \tag{66}$$

and so

$$T_{\text{total}} = T_{\text{trans}} + T_{\text{rot}} + T_{\text{vib}} \tag{67}$$

We shall not consider the effects of vibration–rotation interaction, which are neglected in this first-order treatment. We shall concentrate now on the terms associated with internal vibrations and develop the theory for this term.

3.5. The Potential Energy Function

The potential energy function, V, depends on the coordinates of all of the nuclei in the molecular system under consideration. For the moment, we shall merely put $V = V(R)$, but we shall see in the following sections that to make progress in the "traditional" approach to molecular vibration we have to make some fairly drastic assumptions about the functional form of V.

4. The Secular Equations for Molecular Vibration

4.1. Mass-Adjusted Coordinates

We now consider in more detail the internal motion of the molecular model relative to the moving coordinate system. Recall that $\mathbf{a}_p = (a_p, b_p, c_p)$ are the Cartesian coordinates of the equilibrium position of the atom p and $\boldsymbol{\rho}_p = (\Delta x_p, \Delta y_p, \Delta z_p)$ are the Cartesian displacement coordinates. As a first step in the simplification of this problem, it is useful to define the mass-adjusted displacement coordinates as follows

$$\mathbf{q}_p = (q_{px}, q_{py}, q_{pz}) \tag{68}$$

that is,

$$\mathbf{q}_p = m_p^{1/2}\boldsymbol{\rho}_p \tag{69}$$

We write **q** for the single-column vector of the $3N$ coordinates, q_{px}, in matrix notation, $\mathbf{q}^T$ for its single-row transpose, and q_r, $r = 1, \ldots, 3N$, for a typical element of q_p when we are not interested in the Cartesian components.

The vibrational kinetic energy can be written in terms of the mass-adjusted coordinates as follows

$$\begin{aligned} 2T &= \sum_{p=1}^{N} m_p \dot{\boldsymbol{\rho}}_p^2 \\ &= \sum_{p=1}^{N} \dot{\mathbf{q}}_p^2 \\ &= \sum_{r=1}^{3N} \dot{q}_r^2 \end{aligned} \tag{70}$$

It should be noted that this is a positive definite form.

The potential energy function, V, depends on the coordinates of all the atoms in the molecule. It is usually written as an expansion about the equilibrium position, that is,

$$V = V_0 + \tfrac{1}{2} \sum_r \sum_s f_{rs} q_r q_s + \cdots \tag{71}$$

The constant term in this expansion may be chosen to be zero. The first power term is zero by definition of the equilibrium. If third- and higher-order powers are neglected, then the potential energy function is of the form

$$2V = \sum_{r=1}^{3N} \sum_{s=1}^{N} f_{rs} q_r q_s \tag{72}$$

$$= \mathbf{q}^T \mathbf{f} \mathbf{q} \tag{73}$$

in which

$$f_{rs} = \frac{\partial^2 V}{\partial q_r \, \partial q_s} \tag{74}$$

is an element of a square symmetric matrix **f** of dimension $3N$. This is also a positive definite form because if it were not the molecule would have some unstable coordinate, which we postulate is not the case.

The most convenient form of classical mechanics with which to analyze vibrations is the Lagrangian method. Now, the Lagrangian is

$$L = T - V \tag{75}$$

which, using the forms for the kinetic energy and the potential energy given in equations (70) and (73), respectively, takes the form

$$L = \frac{1}{2}\left(\sum_{r=1} \dot{q}_r^2 - \sum_{r=1}\sum_{s=1} q_r q_s f_{rs}\right) \tag{76}$$

Applying classical mechanics in the form of Lagrange equations

$$\frac{d}{dt}\left(\frac{\partial L}{\partial \dot{q}_r}\right) - \frac{\partial L}{\partial q_r} = 0 \tag{77}$$

we obtain

$$\ddot{q}_r + \sum_{s=1}^{3N} f_{rs} q_s = 0, \qquad r = 1, 2, \ldots, 3N \tag{78}$$

which is a set of $3N$ simultaneous second-order linear differential equations.

We try the solution

$$q_r = A_r \cos(\lambda^{1/2} t + \varepsilon) \tag{79}$$

which gives

$$\sum_s (f_{rs} - \delta_{rs}\lambda) A_s = 0, \qquad r = 1, 2, \ldots, 3N \tag{80}$$

For a nontrivial solution of this set of simultaneous equations, the determinant of the coefficients must vanish, that is,

$$\begin{pmatrix} f_{11} - \lambda & f_{21} & f_{31} & \cdots \\ f_{12} & f_{22} - \lambda & f_{32} & \cdots \\ f_{13} & f_{23} & f_{33} - \lambda & \cdots \\ \vdots & \vdots & \vdots & \end{pmatrix} = 0 \tag{81}$$

which is the secular equation whose $3N$ roots, λ_i, determine the $3N$ possible vibrational frequencies in the coordinates given by (79). Evidently, $\lambda_i^{1/2} = 2\pi\nu_i$ or

$$\lambda_i = 4\pi^2\nu_i^2 \tag{82}$$

where ν_i is the frequency of vibration in the ith normal vibration defined by (79):

$$q_r = A_{ri}\cos(\lambda_i^{1/2}t + \varepsilon) \tag{83}$$

Finally, the relative amplitudes with which the several mass-adjusted Cartesian coordinates q_r vibrate, in a given normal mode λ_i, are obtained by solving (80), that is, the A_{ri}, $r = 1, \ldots, 3N$ are proportional to the cofactors of any one row (or column) of the secular equation with λ_i substituted for λ. It is convenient to normalize the A's such that

$$\sum_{r=1}^{3N} A_{ri} = 1 \tag{84}$$

We note that the $3N$ coordinates, q_r, are not all independent. We have not taken account of the Eckart conditions in setting up the secular equations. There are just $3N-6$ independent internal coordinates, and we have included six zero roots corresponding to overall translation and overall rotation of the molecular model. The Eckart conditions simply require the six corresponding coordinates to be identically zero.

4.2. Normal Coordinates

Another, more powerful, approach to the problem of normal vibrations is through the transformation of the $3N$ coordinates, q_r, to $3N$ new coordinates, Q_i, such that in terms of the new coordinates both the kinetic energy matrix **T** *and* the potential energy matrix **V** are diagonal. Provided that **T** and **V** are positive definite, this is always possible. Moreover, we can always choose either **T** or **V** to have unit coefficients on the diagonal. We shall take **T** to have unit coefficients and thus define the *normal coordinates by*

$$q_r = \sum_{i=1}^{3N} A_{ri} Q_i \tag{85}$$

or in matrix notation

$$\mathbf{q} = \mathbf{A}\mathbf{Q} \tag{86a}$$

$$\mathbf{q}^T = \mathbf{Q}^T\mathbf{A}^T \tag{86b}$$

The kinetic energy may be expressed in terms of normal coordinates as follows

$$2T = \sum_{r=1}^{3N} \dot{q}_r^2 = \sum_{i=1}^{3N} \dot{Q}_i^2 \tag{87}$$

and the potential energy takes the form

$$2V = \sum_{r=1}^{3N} f_{rs} q_r q_s = \sum_{i=1}^{3N} \lambda_i Q_i^2 \tag{88}$$

Using matrix notation, the kinetic energy becomes

$$2T = \dot{\mathbf{q}}^T\dot{\mathbf{q}} = \dot{\mathbf{Q}}^T\mathbf{A}^T\mathbf{A}\dot{\mathbf{Q}} = \dot{\mathbf{Q}}^T\dot{\mathbf{Q}} \tag{89}$$

so that

$$\mathbf{A}^T\mathbf{A} = \mathbf{I} \tag{90}$$

where **I** is the unit matrix of dimension $3N$. Thus, **A** is an orthogonal matrix ($\mathbf{A}^T = \mathbf{A}^{-1}$), and **q** and **Q** are related by an orthogonal transformation.

$$\sum_{i=1}^{3N} A_{ri}A_{si} = \delta_{rs} \tag{91a}$$

$$\sum_{r=1}^{3N} A_{ri}A_{rj} = \delta_{ij} \tag{91b}$$

$$A_{ri} = (A^{-1})_{ir} \tag{91c}$$

In matrix notation, the potential energy takes the form

$$\begin{aligned} 2\mathbf{V} &= \mathbf{q}^T\mathbf{f}\mathbf{q} \\ &= \mathbf{Q}^T\mathbf{A}^T\mathbf{f}\mathbf{A}\mathbf{Q} \end{aligned} \tag{92}$$

so that

$$\mathbf{A}^T\mathbf{f}\mathbf{A} = \boldsymbol{\lambda} \tag{93}$$

where $\boldsymbol{\lambda}$ is a diagonal matrix with diagonal coefficients λ_i.

Equation (93) may be written in the form

$$\mathbf{f}\mathbf{A} = A\lambda \tag{94}$$

or

$$\sum_{s=1}^{3N} f_{rs}A_{si} = \lambda_i A_{ri}, \qquad r = 1, \ldots, 3N \tag{95}$$

which is the same set of simultaneous equations that we obtained above in (80). They have as a condition of self-consistency the equation

$$|\mathbf{f} - \lambda\mathbf{I}| = 0 \tag{96}$$

which is the secular equation (81). When λ is given a value λ_i to satisfy (96), the appropriate column $\mathbf{A}_i$ from the matrix $\mathbf{A}$ is obtained by solving equation (95). The λ's of (95) are identical to the frequency factors of the previous treatment, and the A's to the amplitudes of vibration.

Again, this follows from substituting (87) and (88) into the Lagrange equations. It is apparent then that the problem separates into $3N$ independent simple harmonic motion vibrations, one in each of the $3N$ normal coordinates. The normal coordinates are just the coordinates of the several different normal vibrations, and if just one normal vibration is excited, then all the other normal coordinates remain equal to zero. The total vibrational energy, therefore, is just the sum of the vibrational energies for each of the normal modes. The total vibrational wave function is a product of functions, one for each vibration. In a quantum mechanical treatment these functions are of the form $N_v H_v \exp(-\frac{1}{2}(\omega/\hbar)^{1/2}Q)$, where N_v is a normalization constant, and H_v is a Hermite polynomial. The low-order Hermite polynomials are given in Table 2.

There is no real need to start from mass-adjusted Cartesian coordinates as we have done. (The only particular advantage of these coordinates is that they make the transformation to normal coordinates orthogonal.) We might define

$$2T = \sum_{u,v} \mu_{uv} \dot{q}_u \dot{q}_v \tag{97}$$

$$2V = \sum_{u,v} f_{uv} q_u q_v \tag{98}$$

where the elements μ_{uv} and f_{uv} may be related to those for Cartesian coordinates or any other coordinates once the transformation between coordinate systems has been defined. We then look for normal coordinates Q_i where

$$q_u = \sum_i L_{ui} Q_i \tag{99}$$

such that

$$2T = \sum_i \dot{Q}_i^2 \tag{100}$$

$$2V = \sum_i \lambda_i Q_i^2 \tag{101}$$

Table 2. Hermite Polynomials

$H_0(x) = 1$
$H_1(x) = 2x$
$H_2(x) = 4x^2 - 2$
$H_3(x) = 8x^3 - 12x$
$H_4(x) = 16x^4 - 48x^2 + 12$
$H_5(x) = 32x^5 - 160x^3 + 120x$
$H_6(x) = 64x^6 - 480x^4 + 720x^2 - 120$
$H_7(x) = 128x^7 - 1344x^5 + 3360x^3 - 1680x^2$
$H_8(x) = 256x^8 - 3584x^6 - 13{,}440x^4 - 13{,}440x^2 + 1680$

and in general we find

$$\mathbf{L}^T\boldsymbol{\mu}\mathbf{L} = \mathbf{I} \tag{102}$$

$$\mathbf{L}^T\mathbf{f}\mathbf{L} = \boldsymbol{\lambda} \tag{103}$$

so that

$$\mathbf{f}\mathbf{L} = \boldsymbol{\mu}\mathbf{L}\boldsymbol{\lambda} \tag{104}$$

or for a single column

$$\sum_v f_{uv} L_{vi} = \lambda_i \sum_v \mu_{uv} L_{vi} \tag{105}$$

with the secular equation

$$|f_{uv} - \lambda\mu_{uv}| = 0 \tag{106}$$

It should be noted that in general **L** will not be an orthogonal matrix. However, **f** and **μ** will always be symmetric matrices.

5. Internal Coordinates

Two further developments in the choice of initial coordinates can be very helpful.

One development is to choose a new set of $3N$ coordinates of which six are the external rigid motion of the molecule, which we shall set to zero by the Eckart conditions, and $3N - 6$ are "internal" coordinates.[1] The six roots corresponding to external motion then immediately factor from the secular equation, leaving equations of order $3N - 6$.

The other development is to choose the "internal" coordinates to have the symmetry properties of the molecule, which leads to a further simplification of the secular equation.

5.1. Wilson's S Vectors

We now choose a new set of $3N$ coordinates, $3N - 6$ which describes the internal motion of the molecule,

$$S_u, \qquad u = 1, 2, \ldots, 3N - 6 \tag{107}$$

and six of which describe the rigid motion, translation, and rotation, of the whole molecular model:

$$R_u, \qquad u = 1, 2, \ldots, 6 \tag{108}$$

We choose the R_u's to be

$$\begin{aligned} R_1 &= M^{-1/2} \sum_p m_p^{1/2} q_{xp} \\ &= M^{-1/2} \sum_p m_p x_p \end{aligned} \tag{109}$$

with R_2 and R_3 defined similarly in terms of y and z and

$$\begin{aligned} R_4 &= (I_{xx})^{-1/2} \sum_p m_p^{1/2} (b_p \dot{q}_{zp} - c_p \dot{q}_{yp}) \\ &= (I_{xx})^{-1/2} \sum_p m_p (b_p \dot{z}_p - c_p \dot{y}_p) \end{aligned} \tag{110}$$

R_5 and R_6 are again obtained by permutation of x, y, and z.

We can now choose the S_u's to be defined in terms of the q_t's or the x_t's in any convenient way that is *orthogonal* to the above equations. That is, if

$$S_u = \sum_r D_{ur} q_r \tag{111}$$

$$R_u = \sum_r d_{ur} q_r \tag{112}$$

or in matrix notation

$$\begin{pmatrix} \mathbf{S} \\ \mathbf{R} \end{pmatrix} = \begin{pmatrix} \mathbf{D} & \mathbf{0} \\ \mathbf{0} & \mathbf{d} \end{pmatrix} (\mathbf{q}) \tag{113}$$

then we have already defined the transformation coefficients $\mathbf{d}$ by equations (109) and (110). The only restriction placed on the choice of the matrix $\mathbf{D}$—the definition of the internal coordinates, S_u—is that the matrix

$$\begin{vmatrix} \mathbf{D} & \mathbf{0} \\ \mathbf{0} & \mathbf{d} \end{vmatrix} \tag{114}$$

be orthogonal by rows.

It is now easy to show that a translation of the whole molecular model by a distance Q in the x direction increases the coordinate R_1 by an amount

$$\delta R_1 = M^{-1/2} \sum_t m_t^{1/2}(m_t^{1/2} Q)$$
$$= M^{1/2} Q \tag{115}$$

but leaves every other coordinate unaltered because of the orthogonality.

Similarly, for an infinitesimal rotation θ about the x axis, it can be shown that

$$\delta R_4 = (I_{xx})^{1/2}\theta \tag{116}$$

Thus, $R_1, \ldots, R_6$ do indeed represent rigid motion of the molecular model. Looking again at the Eckart conditions (46) and (52) used to define the rotating axes, we see that they are quite simply

$$R_u = 0, \qquad u = 1, 2, \ldots, 6 \tag{117}$$

The six relations

$$R_u = \sum_r d_{ur} q_r = 0, \qquad u = 1, 2, \ldots, 6 \tag{118}$$

are just the six relations required to reduce the $3N$ coordinates q_r to $3N - 6$ independent coordinates.

Notice also that $R_1, \ldots, R_6$ have been chosen to be mass-adjusted in the sense that the kinetic energy of translation is

$$2T_{\text{trans}} = M(\dot{x}^2 + \dot{y}^2 + \dot{z}^2)$$
$$= \dot{R}_1^2 + \dot{R}_2^2 + \dot{R}_3^2 \tag{119}$$

and the kinetic energy of rotation is

$$2T_{\text{rot}} = I_{xx}\theta_x^2 + I_{yy}\theta_y^2 + I_{zz}\theta_z^2$$
$$= R_4^2 + R_5^2 + R_6^2 \tag{120}$$

We shall find that $R_1, \ldots, R_6$ are actually the six zero-frequency normal coordinates of our treatment above.

We must now obtain the kinetic matrix in the new coordinates. We denote the transformation matrices

$$\mathbf{S} = \mathbf{Dq} \tag{121}$$

and

$$\mathbf{R} = \mathbf{dq} \tag{122}$$

or

$$\mathbf{S} = \mathbf{Bx} \tag{123}$$

and

$$\mathbf{R} = \mathbf{bx} \tag{124}$$

that is,

$$\begin{pmatrix}\mathbf{S}\\ \mathbf{R}\end{pmatrix} = \begin{pmatrix}\mathbf{D} & \mathbf{0}\\ \mathbf{0} & \mathbf{d}\end{pmatrix}(\mathbf{q}) = \begin{pmatrix}\mathbf{B} & \mathbf{0}\\ \mathbf{0} & \mathbf{b}\end{pmatrix}(\mathbf{x}) \tag{125}$$

and evidently

$$B_{ur} = m_r^{1/2} D_{ur} \qquad \mathbf{B} = M^{1/2}\mathbf{D} \tag{126}$$

$$b_{ur} = m_r^{1/2} d_{ur} \tag{127}$$

$$\mathbf{b} = M^{1/2}\mathbf{d} \tag{128}$$

For the inverse transformation

$$(\mathbf{q}) = \begin{pmatrix}\mathbf{C} & \mathbf{0}\\ \mathbf{0} & \mathbf{c}\end{pmatrix}\begin{pmatrix}\mathbf{S}\\ \mathbf{R}\end{pmatrix} \tag{129}$$

$$(\mathbf{x}) = \begin{pmatrix}\mathbf{A} & \mathbf{0}\\ \mathbf{0} & \mathbf{a}\end{pmatrix}\begin{pmatrix}\mathbf{S}\\ \mathbf{R}\end{pmatrix} \tag{130}$$

Evidently, the matrix **C** is the inverse of **D**, and **B**, similarly, is the inverse of **A**.

Now we can obtain the vibrational kinetic energy in the form

$$2T = \dot{\mathbf{S}}^T(\mathbf{G})^{-1}\dot{\mathbf{S}} \tag{131}$$

where

$$\mathbf{G} = \mathbf{B}\mathbf{M}^{-1}\mathbf{B}^T \tag{132}$$

so that the matrix **G** contains information both about the atomic masses (**M**) and the equilibrium geometry of the molecule (**B**). The potential energy may be written

$$2V = \mathbf{S}^T\mathbf{F}\mathbf{S} \tag{133}$$

where

$$\mathbf{F} = \mathbf{A}^T\mathbf{f}\mathbf{A} \tag{134}$$

The vibrational eigenproblem then assumes the "FG form":

$$\mathbf{G}\mathbf{F}\boldsymbol{\Phi} = \boldsymbol{\Phi}\boldsymbol{\lambda} \tag{135}$$

which has solutions for which

$$|\mathbf{G}\mathbf{F} - \lambda\mathbf{I}| = 0 \tag{136}$$

5.2. Point Group Symmetry

We illustrate the use of point group symmetry in molecular vibration calculations by considering a simple example, H_2O. The vibration motion of the water molecule may be described in terms of the internal coordinates

$$\mathbf{R} = \begin{pmatrix} \delta r_1 \\ \delta r_2 \\ r_0\delta\alpha \end{pmatrix} \tag{137}$$

where δr_1 and δr_2 are associated with stretching of the O—H bonds and $r_0\delta\alpha$ with bond angle bending. This molecule has C_{2v} point group symmetry.

Table 3. Character Table for the C_{2v} Point Symmetry Group

C_{2v}	E	C_2	$\sigma_v(xz)$	$\sigma_v'(yz)$	
A_1	1	1	1	1	z
A_2	1	1	−1	−1	R_z
B_1	1	−1	1	−1	xR_y
B_2	1	−1	−1	1	yR_x

The character table for this point group is shown in Table 3. We may transform the internal coordinates to symmetry coordinates by means of the transformation

$$\begin{pmatrix} S_1 \\ S_2 \\ S_3 \end{pmatrix} = \begin{pmatrix} (2)^{-1/2} & (2)^{-1/2} & 0 \\ 0 & 0 & 1 \\ (2)^{-1/2} & -(2)^{-1/2} & 0 \end{pmatrix} \begin{pmatrix} \delta r_1 \\ \delta r_2 \\ r_0 \delta \alpha \end{pmatrix} \tag{138}$$

where S_1 is the symmetric stretching motion, S_2 is the antisymmetric stretching motion, and S_3 is the bending motion shown in Fig. 3.

The kinetic energy matrix may be written in terms of the internal coordinates as follows:

$$\mathbf{G} = \begin{pmatrix} (\mu_H + \mu_O) & \mu_O \cos\alpha & -\mu_O \sin\alpha \\ \mu_O \cos\alpha & (\mu_H + \mu_O) & -\mu_O \sin\alpha \\ -\mu_O \sin\alpha & -\mu_O \sin\alpha & \mu_H + \mu_O(1 - \cos\alpha) \end{pmatrix} \tag{139}$$

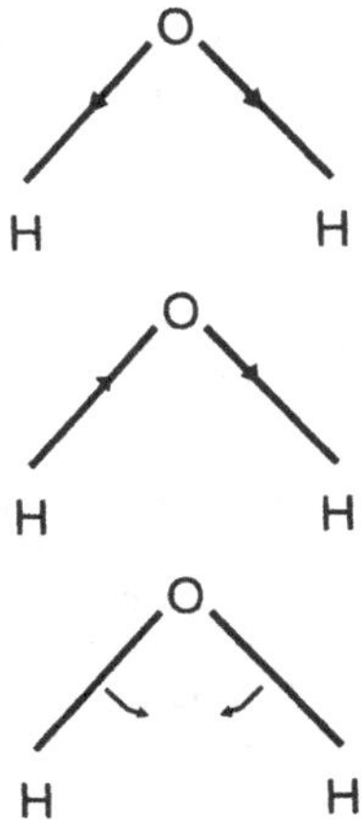

Figure 3. The symmetric stretching motion, the antisymmetric stretching motion, and the bending motion in the water molecule.

By transforming to symmetry coordinates the kinetic energy matrix may be cast in the form

$$\mathbf{G} = \begin{pmatrix} \mu_H + \mu_O(1 + \cos\alpha) & -\mu_O(2)^{1/2}\sin\alpha & 0 \\ \mu_O(2)^{1/2}\sin\alpha & 2[\mu_H + \mu_O(1 - \cos\alpha)] & 0 \\ 0 & & \mu_H + \mu_O(1 - \cos\alpha) \end{pmatrix} \tag{140}$$

If we employ a simple valence force field potential

$$V = \tfrac{1}{2}f_{11}S_1^2 + \tfrac{1}{2}f_{22}S_2^2 + \tfrac{1}{2}f_{33}S_3^2 \tag{141}$$

then the **F** matrix assumes a diagonal form, and for this simple problem the "FG form" is factored into a 2 × 2 problem and a one-dimensional problem.

6. Concluding Remarks

The purpose of this chapter has been to provide an overview of the "traditional" approach to the problem of describing vibration in molecules. This approach, which was developed in the precomputer era, aims to simplify the problem as much as possible so as to render it tractable to hand calculation.

The "traditional" approach runs into difficulties when, for example, the vibrational amplitude is not small. The quadratic force field is usually not adequate in such situations. Difficulties also arise when there is significant coupling between vibrational and rotational motion or between vibrational and electronic motion.

Our aim in this chapter has been to provide a background for the more advanced contributions that follow.

References

1. E. B. Wilson, J. C. Decius, and P. C. Cross, *Molecular Vibrations: The Theory of Infrared and Raman Spectra*, McGraw-Hill, New York (1955).
2. G. Herzberg, *Molecular Spectra and Molecular Structure. II. Infrared and Raman Spectra of Polyatomic Molecules*, Van Nostrand, New York (1945).
3. J. W. Cooley, *Math. Comp.* **15**, 363 (1961).
4. J. K. Cashion, *J. Chem. Phys.* **39**, 1872 (1963).
5. J. L. Dunham, *Phys. Rev.* **41**, 721 (1932).
6. P. F. Fougere and R. K. Nesbet, *J. Chem. Phys.* **44**, 285 (1966).

7. G. Simons, R. J. Parr, and J. M. Finlan, *J. Chem. Phys.* **59**, 3229 (1973).
8. C. L. Beckel, *J. Chem. Phys.* **65**, 4319 (1976).
9. S. Wilson, *Molec. Phys.* **35**, 1 (1978).
10. S. Wilson, *Electron Correlation in Molecules*, Clarendon Press, Oxford (1984).
11. M. Born and J. R. Oppenheimer, *Ann. Phys. Leipzig* **84**, 457 (1927).
12. M. Born and K. Huang, *Dynamical Theory of Crystal Lattices*, Oxford University Press, Oxford, England (1954).
13. C. Eckart, *Phys. Rev.* **47**, 52 (1935).
14. J. Louck and H. Galbraith, *Rev. Mod. Phys.* **48**, 69 (1976).

2

The Mathematics of Vibration–Rotation Calculations

BRIAN T. SUTCLIFFE

1. Introduction

Since the very beginning of the application of quantum mechanics to molecules,[1,2] it has been assumed that a complete calculation should be undertaken in two stages. In the first stage, the nuclear positions are to be treated as parameters in a quantum-mechanical electronic structure calculation. In the second stage, the results of the electronic structure calculation are to be used to generate a potential in which the nuclear motion calculation is to be performed. Thus, the electronic and nuclear parts of the problem are decoupled.

It is widely believed that this approach is justified by reference to work presented in a paper by Born and Oppenheimer[3] published in 1927.

In practice, however, very little use has been made of the analytical apparatus of that paper to justify in detail what was actually being done in calculations—or, to be a little more truthful, what it was supposed could be done in calculations—for until the advent of digital computers detailed calculation often was simply too difficult to be contemplated. The paper by Born and Oppenheimer has been subject to reexamination in more recent

BRIAN T. SUTCLIFFE • Department of Chemistry, University of York, York YO1 5DD, U.K.

Methods in Computational Chemistry, Volume 4: Molecular Vibrations, edited by Stephen Wilson, Plenum Press, New York, 1992.

times,[4,5] and this more recent work perhaps explains why the actual approach used by Born and Oppenheimer never influenced to any extent the formulation of practical calculations. Nevertheless, the decoupling idea persisted.

It was assumed that conventional clamped-nucleus electronic structure calculations could be performed at different nuclear geometries to yield an electronic energy at each geometry. The nuclear repulsion calculated classically at each geometry was then to be added to the electronic energy, and the energies found in this way were then to be regarded as representative values of a smoothly varying function called the potential energy function or surface, evaluated at the particular points. Because the points chosen for the different calculations correspond simply to different geometries, the potential depends only on the geometry of the nuclear configuration and is invariant under uniform translations or rigid rotations. For the ground states of all stable molecules it was assumed that there would be a deep minimum on the potential energy surface at a particular nuclear geometry, the equilibrium geometry, and this geometry was thought of as defining the shape of the molecule. It is this idea of an invariant potential and of the equilibrium geometry to which a minimum in this potential gives rise that Eckart[6] took as the basis of his seminal discussion of molecular vibration–rotation motion. Eckart considered just the nuclei in the problem and took the potential as a given. In terms of these nuclear variables he developed the Hamiltonian that provides the basic model for interpreting and fitting almost all molecular spectra. He did not put his Hamiltonian into quantum-mechanical form, and indeed a completely satisfactory quantum-mechanical form was only achieved in the work of Watson[7] published in 1968.

The Eckart–Watson Hamiltonian was used principally as a device for assigning experimental results in molecular spectra studies on the basis of an assumed perturbation theory approach. Recently, however, attempts have been made to predict molecular spectra by ab initio calculations using this Hamiltonian with a computed potential function[8,9] obtained from electronic structure calculations using the standard clamped-nucleus Hamiltonian. It is this and related work, especially on triatomic systems, that after 50 years has stimulated interest again, both in different possible forms for a nuclear motion Hamiltonian and in the precise relationship between the nuclear motion problem and the full problem. It is these topics to which this review will be directed.

An approach to the separation of electronic and nuclear motion which seemed to offer more help in practical calculations than that of the original work of Born and Oppenheimer was presented first by Born in 1951 and written up in a generally available form in Born and Huang[10] in 1954. This approach is usually assumed to supersede and subsume the older one, and most popular expositions use this approach and call it the Born–Oppenheimer approximation. (It is interesting to note that the same idea had actually been proposed by Slater[11] somewhat earlier than the original Born and Oppenheimer paper.) There have been a small number of pioneering calculations on diatomic systems (see Ref. 12, for example) that have utilized this approach in practice, and it is this approach that will inform the present review.

In the Born and Huang approach, the wave function for the full problem is expressed in terms of products of eigenstates of an electronic Hamiltonian with functions of the nuclear variables only. The functions of the nuclear variables are determined by solving a problem posed in terms of the potential energy surfaces associated with electronic eigenstates. Because the potentials are translation–rotation invariant, it is clear that they must arise from the solution of a body-fixed problem from which translations and rotations have been separated. How such a body-fixed problem should be constructed is the subject of the next section. In the section that follows that, the nature of the product function separation is considered.

2. The Full Molecule Problem

It will be assumed both that a molecule can be described by the ordinary Schrödinger form of the Hamiltonian for the appropriate number of electrons and nuclei and that the dominant interactions between the particles is the Coulomb one. Spin will be considered only insofar as it determines the statistics of the particles.

Consider a collection of N particles labeled in the laboratory-fixed frame as x_i, $i = 1, 2, \ldots, N$ with masses m_i and charges Z_ie. The charge numbers Z_i are positive for a nucleus and minus one for an electron. In a neutral system the charge numbers sum to zero. It will be convenient to think of the $\mathbf{x}_i$ as a column matrix of three Cartesian components $x_{\alpha i}$, $\alpha = x, y, z$ and to think of the $\mathbf{x}_i$ collectively as the $3 \times N$ matrix $\mathbf{x}$.

The separation between particles is then defined by

$$x_{ij}^2 = \sum_{\alpha} (x_{\alpha j} - x_{\alpha i})^2 \tag{1}$$

where the alpha sum runs over x, y, and z.

The laboratory-fixed form of the Schrödinger Hamiltonian describing this system of N particles is then

$$\hat{H}(\mathbf{x}) = -\frac{\hbar^2}{2}\sum_{i=1}^{N} m_i^{-1}\nabla^2(\mathbf{x}_i) + \frac{e^2}{8\pi\varepsilon_0}\sum_{i,j=1}^{N}{}' \frac{Z_iZ_j}{x_{ij}} \tag{2}$$

where the notation is standard.

It is easy to see that the Hamiltonian operator is invariant under the following operations:

1. Uniform translations $\mathbf{x}_i \to \mathbf{x}_i + \mathbf{a}$.
2. Orthogonal transformations $\mathbf{x}_i \to \mathbf{R}\mathbf{x}_i$, where $\mathbf{R}$ is a 3×3 orthogonal matrix such that $\mathbf{R}^T\mathbf{R} = \mathbf{E}_3$, $|\mathbf{R}| = \pm 1$.
3. Permutations of identical particles, that is, $\mathbf{x}_i \to \mathbf{x}_j$ if $m_i = m_j$ and $Z_i = Z_j$.

Each of these sets of operations forms a group, so it is clearly possible to classify the eigenfunctions of $\hat{H}$ according to the characteristics of these groups. More particularly, it can be shown that the properties of the first group allow the Hamiltonian to be split into two parts, one that carries all the translation motion and the other (the space-fixed part) that is expressed in terms of translationally invariant coordinates. Because an orthogonal transformation in three dimensions represents either a proper or improper rotation (this last, a rotation-inversion or reflection), the properties of the second group allow the space-fixed Hamiltonian to be expressed in terms of rotational operators and operators that depend on variables (internal coordinates) that are invariant to orthogonal transformations. This is the body-fixed Hamiltonian. It is further possible to specify, at least formally, the rotational eigenfunctions of the body-fixed problem and hence to replace the full problem with an effective Hamiltonian expressible in terms of the internal variables only, by integrating out over the rotations. Because the solutions of this effective Hamiltonian are rotationally and translationally invariant, it is here that the origin of the potential surface must reside. The approach used by Eckart was in terms of a particular choice made for the body-fixed Hamiltonian, but for present purposes an attempt is made to keep the discussion as general as possible. Permutational symmetry requirements are not immediately relevant to the way in which a body-fixed Hamiltonian is constructed, so consideration of this symmetry will be deferred until later.

First consider the space-fixed Hamiltonian. This may be constructed by the coordinate transformation symbolized by:

$$(\mathbf{t}\mathbf{X}_T) = \mathbf{x}\mathbf{V} \tag{3}$$

In (3), $\mathbf{t}$ is a $3 \times N-1$ matrix, and $\mathbf{X}_T$ is a 3×1 matrix. $\mathbf{V}$ is an $N \times N$ matrix which, from the structure of the left-hand side of (3), has a special last column whose elements are

$$V_{iN} = M_T^{-1} m_i, \qquad M_T = \sum_{i=1}^{N} m_i \tag{4}$$

so that $\mathbf{X}_T$ is the standard center-of-mass coordinate:

$$\mathbf{X}_T = M_T^{-1} \sum_{i=1}^{N} m_i x_i \tag{5}$$

The coordinates $\mathbf{t}_j$, $j = 1, 2, \ldots, N-1$ are to be translationally invariant, so it is required on each of the remaining columns of $\mathbf{V}$ that

$$\sum_{i=1}^{N} V_{ij} = 0, \qquad j = 1, 2, \ldots, N-1 \tag{6}$$

and it is easy to see that (6) forces $\mathbf{t}_j \to \mathbf{t}_j$ as $\mathbf{x}_i \to \mathbf{x}_i + \mathbf{a}$, all i.

The $\mathbf{t}_i$ are independent if the inverse transformation

$$\mathbf{x} = (\mathbf{t}\mathbf{X}_T)\mathbf{V}^{-1} \tag{7}$$

exists. The structure of the right-hand side of (7) shows that the bottom row of $\mathbf{V}^{-1}$ is special, and it is easy to see that, without loss of generality, we may require its elements to be

$$(\mathbf{V}^{-1})_{Ni} = 1, \qquad i = 1, 2, \ldots, N \tag{8}$$

The inverse requirement on the remainder of $\mathbf{V}^{-1}$ implies that

$$\sum_{i=1}^{N} (\mathbf{V}^{-1})_{ji} m_i = 0, \qquad j = 1, 2, \ldots, N-1 \tag{9}$$

If we write the column matrix of the Cartesian components of the partial derivative operator as $\partial/\partial\mathbf{x}_i$, then the coordinates change (3) gives:

$$\frac{\partial}{\partial \mathbf{x}_i} = m_i M_T^{-1} \frac{\partial}{\partial \mathbf{X}_T} + \sum_{j=1}^{N-1} V_{ij} \frac{\partial}{\partial \mathbf{t}_j} \tag{10}$$

and hence the Hamiltonian (2) in the new coordinates becomes

$$\hat{H}(\mathbf{t}, \mathbf{X}_T) = -\frac{\hbar^2}{2M_T}\nabla^2(\mathbf{X}_T) - \frac{\hbar^2}{2}\sum_{i,j=1}^{N-1} \mu_{ij}^{-1}\nabla(\mathbf{t}_i)\cdot\nabla(\mathbf{t}_j)$$
$$+\frac{e^2}{8\pi\varepsilon_0}\sum_{i,j=1}^{N}{}' \frac{Z_i Z_j}{f_{ij}(\mathbf{t})} \tag{11}$$

Here

$$\mu_{ij}^{-1} = \sum_{k=1}^{N} m_k^{-1} V_{ki} V_{kj}, \qquad i, j = 1, 2, \ldots, N-1 \tag{12}$$

and f_{ij} is just x_{ij} as given by (1) but expressed as a function of the $\mathbf{t}_i$. Thus,

$$f_{ij}(\mathbf{t}) = \left(\sum_{\alpha}\left\{\sum_{k=1}^{N-1}[(\mathbf{V}^{-1})_{kj} - (\mathbf{V}^{-1})_{ki}]t_{\alpha k}\right\}^2\right)^{1/2} \tag{13}$$

In (11), the $\nabla(\mathbf{t}_i)$ are the usual grad operators expressed in the Cartesian components of $\mathbf{t}_i$, and the first term represents the center-of-mass kinetic energy. Because the center-of-mass variable does not enter the potential term, the center-of-mass problem may be separated completely so that the full solution is of the form

$$T(\mathbf{X}_T)\Psi(\mathbf{t}) \tag{14}$$

where

$$T(\mathbf{X}_T) = \exp(i\mathbf{k}\mathbf{X}_T), \qquad \mathbf{k} \equiv (k_x, k_y, k_z) \tag{15}$$

and where the associated translational energy is

$$E_T = \frac{|\mathbf{k}|^2}{2M_T} \tag{16}$$

It should be noted that the translational wave function is not square integrable and that the translational energy is continuous. This is exactly what is to be expected given that the group of translations in three dimensions is a noncompact continuous group and has no finite-dimension irreducible representations. It should also be noted that it is absolutely essential that the translational motion be separated from the problem before any approximate solution is attempted in order to avoid the continuous spectrum, which cannot be approximated. Thus, attention will be confined to the remaining terms in (11), which will be denoted collectively by $\hat{H}(\mathbf{t})$ and referred to as the space-fixed Hamiltonian. In the space-fixed Hamiltonian, the inverse effective mass matrix $\boldsymbol{\mu}^{-1}$ and the form of the potential functions f_{ij} depend intimately on the choice of $\mathbf{V}$, and the choice of this is essentially arbitrary. In particular, it should be observed that because there are only $N-1$ space-fixed variables, they cannot, except in the most conventional of senses, be thought of as particle coordinates. The nondiagonal nature of $\boldsymbol{\mu}^{-1}$ and the peculiar form of the f_{ij} also militate against any simple particle interpretation of the space-fixed Hamiltonian.

Nevertheless, rather delicate mathematical considerations (see Ref. 13, for example) indicate that for neutral and positively charged systems the space-fixed Hamiltonian as defined above does have square integrable eigenfunctions among its solutions and also that its spectrum is invariant under different choices of $\mathbf{V}$.

For later purposes it is convenient to have available both the angular momentum operator and the dipole-length operator in terms of X_T and the $\mathbf{t}_i$. The total angular momentum operator may be written as

$$\hat{\mathbf{L}}(\mathbf{x}) = \frac{\hbar}{i} \sum_{i=1}^{N} \hat{\mathbf{x}}_i \frac{\partial}{\partial \mathbf{x}_i} \tag{17}$$

where $\hat{\mathbf{L}}(\mathbf{x})$ are column matrices of Cartesian components, and the skew-symmetric matrix $\hat{\mathbf{x}}_i$ is

$$\hat{\mathbf{x}}_i = \begin{pmatrix} 0 & -x_{zi} & x_{yi} \\ x_{zi} & 0 & -x_{xi} \\ -x_{yi} & x_{xi} & 0 \end{pmatrix} \tag{18}$$

The matrix $\hat{x}_i$ can also be written in terms of the infinitesimal rotation generators:

$$\mathbf{M}^x = \begin{pmatrix} 0 & 0 & 0 \\ 0 & 0 & 1 \\ 0 & -1 & 0 \end{pmatrix} \quad \mathbf{M}^y = \begin{pmatrix} 0 & 0 & -1 \\ 0 & 0 & 0 \\ 1 & 0 & 0 \end{pmatrix} \quad \mathbf{M}^z = \begin{pmatrix} 0 & 1 & 0 \\ -1 & 0 & 0 \\ 0 & 0 & 0 \end{pmatrix} \tag{19}$$

so that

$$\hat{\mathbf{x}}_i = \sum_{\alpha} x_{\alpha i} \mathbf{M}^{\alpha T} \tag{20}$$

A variable symbol with a caret over it will from now on be used to denote a skew-symmetric matrix as defined by (20).

Transforming to the coordinate $\mathbf{X}_T$, $\mathbf{t}_i$ gives

$$\hat{\mathbf{L}}(\mathbf{x}) \rightarrow \frac{\hbar}{i} \hat{\mathbf{X}}_t \frac{\partial}{\partial \mathbf{X}_t} + \frac{\hbar}{i} \sum_{i=1}^{N-1} \hat{\mathbf{t}}_i \frac{\partial}{\partial \mathbf{t}_i} \tag{21}$$

and in later sections the second term will be denoted as $\hat{\mathbf{L}}(\mathbf{t})$ and called the space-fixed angular momentum.

The total dipole operator may be written as:

$$\mathbf{d}(\mathbf{x}) = e \sum_{i=1}^{N} Z_i \mathbf{x}_i \tag{22}$$

and simple transformation using (7) leads to

$$\mathbf{d}(\mathbf{t}, \mathbf{X}_T) = e \sum_{i=1}^{N-1} \tilde{Z}_i \mathbf{t}_i + e Z_T \mathbf{X}_T \tag{23}$$

where the first term in (23) will be denoted $\mathbf{d}(\mathbf{t})$ and the effective charges are given by

$$\tilde{Z}_i = \sum_{j=1}^{N} (\mathbf{V}^{-1})_{ij} Z_j, \qquad Z_T = \sum_{i=1}^{N} Z_i \tag{24}$$

As is to be expected, the center-of-mass dependent term in the dipole vanishes if the system is neutral, that is, if $Z_T = 0$.

To move from the space-fixed system to a body-fixed system, three variables are introduced which define the orientation of the system, together with $3N - 6$ variables which describe the internal motions of the system. The internal variables are chosen to be invariant to orthogonal transformations. To construct the body-fixed system it is supposed that the three orientation variables are specified by means of an orthogonal matrix $\mathbf{C}$ which can be parameterized by the three Euler angles ϕ_m, $m = 1, 2, 3$ as orientation variables. (It should be noted that for $N = 2$ only two orientation variables are required; this rather special case will be ignored in what follows.) Thus, the space-fixed Cartesian coordinates $\mathbf{t}$ can be thought of as being related to a body-fixed set $\mathbf{z}$ by

$$\mathbf{t} = \mathbf{C}\mathbf{z} \tag{25}$$

The above equation defines the variables $\mathbf{z}_i$ so that

$$\mathbf{z}_i = \mathbf{C}^T\mathbf{t}_i \tag{26}$$

and thus any orthogonal transformation of the $\mathbf{t}_i$ leaves the $\mathbf{z}_i$ invariant. However, not all the $3N - 3$ components of the $\mathbf{z}_i$ are independent, for there must be three relations between them. What this means is that the components of $\mathbf{z}_i$ must be expressible in terms of $3N - 6$ independent internal coordinates q_i, $i = 1, 2, \ldots, 3N - 6$. Of course, it is possible that some of the q_i are some of the components of the $\mathbf{z}_i$, but generally speaking the q_i will be expressible in terms of scalar products of the $\mathbf{t}_i$ (and equally of the $\mathbf{z}_i$) because scalar products are the most general constructions that are invariant under orthogonal transformations of their constituent vectors.*

It should be made explicit here that a particular choice of $\mathbf{C}$ does not force a particular choice of $\mathbf{q}$ and that different choices are possible for any given $\mathbf{C}$. Conversely, different choices of $\mathbf{C}$ are possible for any choice of $\mathbf{q}$. Some concrete examples of this sort of freedom of choice are given in Ref. 14.

To express the space-fixed differential operators in body-fixed terms it is necessary to obtain expressions for the partial derivatives of the body-fixed variables with respect to the space-fixed ones. To deal with the angular part first note that

$$\frac{\partial}{\partial t_{\alpha i}}(\mathbf{C}^T\mathbf{C}) = \mathbf{0}_3 \tag{27}$$

because $\mathbf{C}$ is an orthogonal matrix and hence $\mathbf{C}^T\mathbf{C} = \mathbf{E}_3$.

From this it follows at once that

$$\frac{\partial \mathbf{C}^T}{\partial t_{\alpha i}}\mathbf{C} = \hat{\boldsymbol{\omega}}^{\alpha i} \tag{28}$$

where $\hat{\boldsymbol{\omega}}^{\alpha i}$ is a skew-symmetric matrix of the same form as (18), containing

* If only proper orthogonal transformations are considered the scalar triple products are also invariants, but they change sign under improper operations. This and related matters are discussed in Appendix 2 at the end of this chapter.

three independent elements $\omega_\gamma^{\alpha i}$. Using the form of (20), (28) can be rewritten as

$$\frac{\partial \mathbf{C}}{\partial t_{\alpha i}} = \sum_\gamma \omega_\gamma^{\alpha i} \mathbf{C}\mathbf{M}^\gamma \tag{29}$$

It is also convenient to introduce the matrix with elements $\Omega_{\beta\gamma}^i$ such that

$$\omega_\gamma^{\alpha i} = \sum_\beta C_{\alpha\beta} \Omega_{\beta\gamma}^i \tag{30}$$

This ensures that the elements of the matrix $\mathbf{\Omega}^i$ are functions of the internal coordinates only, and thus (29) becomes

$$\frac{\partial \mathbf{C}}{\partial t_{\alpha i}} = \sum_\beta (\mathbf{C}\mathbf{M}^\beta)(\mathbf{C}\mathbf{\Omega}^i)_{\alpha\beta} \tag{31}$$

Recognizing that $\mathbf{C}$ is a function of the ϕ_m only, it follows that

$$\frac{\partial \mathbf{C}}{\partial t_{\alpha i}} = \sum_{m=1}^{3} \frac{\partial \mathbf{C}}{\partial \phi_m} \frac{\partial \phi_m}{\partial t_{\alpha i}} \tag{32}$$

By comparison of (27) to (29) it follows that

$$\frac{\partial \mathbf{C}}{\partial \phi_m} = \sum_\gamma (\mathbf{D}^{-1})_{m\gamma} \mathbf{C}\mathbf{M}^\gamma \tag{33}$$

where $(\mathbf{D}^{-1})_{m\gamma}$ plays the same role in (33) that $\omega_\gamma^{\alpha i}$ does in (29).

From (31) to (33) it follows that

$$\sum_{m=1}^{3} (\mathbf{D}^{-1})_{m\gamma} \frac{\partial \phi_m}{\partial t_{\alpha i}} = (\mathbf{C}\mathbf{\Omega}^i)_{\alpha\gamma}$$

or

$$\frac{\partial \phi_m}{\partial t_{\alpha i}} = (\mathbf{C}\mathbf{\Omega}^i\mathbf{D})_{\alpha m} \tag{34}$$

Of course, the process so far is purely formal because **C** has not been specified in terms of the $\mathbf{t}_i$ or the ϕ_m. The process is not vacuous, however, as will be seen shortly. A similar formal process enables it to be established that

$$\frac{\partial q_k}{\partial t_{\alpha i}} = (\mathbf{CQ}^i)_{\alpha k} \tag{35}$$

where the elements of $\mathbf{Q}^i$ can be shown to depend on internal variables only using the fact that the q_k can be written as functions of the scalar products of the $\mathbf{t}_i$. Thus, it is possible to write

$$\frac{\partial}{\partial \mathbf{t}_i} = \mathbf{C}\left(\mathbf{\Omega}^i\mathbf{D}\frac{\partial}{\partial \boldsymbol{\phi}} + \mathbf{Q}^i\frac{\partial}{\partial \mathbf{q}}\right) \tag{36}$$

where $\partial/\partial\boldsymbol{\phi}$ and $\partial/\partial\mathbf{q}$ are column matrices of 3 and $3N - 6$ partial derivatives, respectively.

It is clear that (34) and (35) are expressions of the Jacobian matrix elements for the transformation from the $(\boldsymbol{\phi}\mathbf{q})$ to the $\mathbf{t}$. It is useful to have expressions for the inverse transformations, and these may, again formally, be obtained from (25). Thus, because the $\mathbf{z}$ cannot be functions of the $\boldsymbol{\phi}$,

$$\frac{\partial t_{\alpha i}}{\partial \phi_m} = \left(\frac{\partial \mathbf{C}}{\partial \phi_m}\mathbf{z}_i\right)_\alpha \tag{37}$$

and using (33) it follows that

$$\frac{\partial t_{\alpha i}}{\partial \phi_m} = (\mathbf{C}\hat{\mathbf{z}}_i\mathbf{D}^{-\mathrm{T}})_{\alpha m} \tag{38}$$

Again using (25) it follows in a way analogous to (35) that

$$\frac{\partial t_{\alpha i}}{\partial q_k} = (\mathbf{C}\tilde{\mathbf{Q}}^i)_{\alpha k} \tag{39}$$

and it can be shown that $\tilde{\mathbf{Q}}^i$ has elements that depend only on the q_k.

The relationship between the Jacobian matrix and its inverse leads to the following expressions:

$$\sum_{i=1}^{N-1} \hat{\mathbf{z}}_i^T \mathbf{\Omega}^i = \mathbf{E}_3, \qquad \sum_{i=1}^{N-1} \tilde{\mathbf{Q}}^{iT} \mathbf{Q}^i = \mathbf{E}_{3N-6}$$

$$\sum_{i=1}^{N-1} \hat{\mathbf{z}}_i^T \mathbf{Q}^i = \mathbf{0}_{3,3N-6}, \qquad \sum_{i=1}^{N-1} \tilde{\mathbf{Q}}^{iT} \mathbf{\Omega}^i = \mathbf{0}_{3N-6,3} \tag{40}$$

and

$$\mathbf{\Omega}^i \hat{\mathbf{z}}_j^T + \mathbf{Q}^i \tilde{\mathbf{Q}}^{jT} = \delta_{ij} \mathbf{E}_3 \tag{41}$$

These expressions are helpful in the formal manipulations that lead to body-fixed forms for the operators. They are also the origin of the "sum rules" that constitute such a part of the manipulation of the Eckart Hamiltonian (see Ref. 7, for example).

Finally note that on the formal variable change (25) it follows that

$$\hat{\mathbf{t}}_i = |\mathbf{C}|\mathbf{C}\hat{\mathbf{z}}_i\mathbf{C}^T \tag{42}$$

where $|\mathbf{C}|$ is either plus or minus one according to whether $\mathbf{C}$ corresponds to a proper rotation or to an improper rotation.

It is convenient first to derive a body-fixed expression for the space-fixed angular momentum operator as given by the second term on the right in (21). Using this, (36) and (42), together with (40), yields

$$\hat{\mathbf{L}}(\mathbf{t}) = -\frac{\hbar}{i} |\mathbf{C}|\mathbf{C}\mathbf{D} \frac{\partial}{\partial \boldsymbol{\phi}} \tag{43}$$

At this stage there is an element of choice for the definition of the body-fixed angular momentum. In this work it will be chosen as

$$\hat{\mathbf{L}}(\boldsymbol{\phi}) = \frac{\hbar}{i} \mathbf{D} \frac{\partial}{\partial \boldsymbol{\phi}} \tag{44}$$

and with the aid of (33) it is possible to show that the components of the operators with this choice obey the standard commutation conditions. It is somewhat more usual to choose the negative of (44) as the definition, in which case its components obey the celebrated anomalous commutation conditions. Whatever the choice, however, the components of the body-fixed angular momentum operator are functions of the ϕ_m alone. With the present choice it may be imagined that there are angular momentum eigenfunctions $|JMk\rangle$, just as in Brink and Satchler[15] or in Biedenharn and Louck,[16] such that

$$
\begin{aligned}
\hat{L}^2(\mathbf{t})|JMk\rangle &= \hat{L}^2(\boldsymbol{\phi})|JMk\rangle = \hbar^2 J(J+1)|JMk\rangle \\
\hat{L}_z(\mathbf{t})|JMk\rangle &= \hbar M|JMk\rangle \\
\hat{L}_z(\boldsymbol{\phi})|JMk\rangle &= \hbar k|JMk\rangle
\end{aligned}
\tag{45}
$$

Furthermore, defining the step-up and down operators in the usual way as $L_\pm = L_x \pm iL_y$, then

$$
\begin{aligned}
\hat{L}_\pm(\mathbf{t})|JMk\rangle &= \hbar C^\pm_{JM}|JM\pm 1k\rangle \\
\hat{L}_\pm(\boldsymbol{\phi})|JMk\rangle &= \hbar C^\mp_{Jk}|JMk\pm 1\rangle
\end{aligned}
\tag{46}
$$

where the phase conventions are chosen analogously to the standard Condon and Shortley ones (see Ref. 15, for example) so that

$$C^\pm_{Jj} = [J(J+1) - j(j\pm 1)]^{1/2} \tag{47}$$

If $\mathbf{C}$ is parameterized by the standard Euler angle choice made in Ref. (15) or (16), then it is easy to show that

$$|JMk\rangle = \left(\frac{2J+1}{8\pi^2}\right)^{1/2} (-1)^k \mathscr{D}^{J*}_{M-k}(\boldsymbol{\phi}) \tag{48}$$

where $\mathscr{D}^J$ is the standard Wigner matrix as defined in Ref. (15) or (16). The functions $|JMk\rangle$ are often called symmetric-top eigenfunctions. Had the more usual choice of the negative of (44) been made for the body-fixed angular momentum operator, the symmetric-top functions would have involved $\mathscr{D}^{J*}_{Mk}$, but in that case to get the second expression in (46) it would have been necessary to redefine the step-up and -down operators as $\hat{L}^\pm(\boldsymbol{\phi}) = \hat{L}_x(\boldsymbol{\phi}) \mp i\hat{L}_y(\boldsymbol{\phi})$. A more extended discussion of these matters can be found in Section 3.8 of Ref. (16).

In fact, it can be shown[16,17] that, whatever the parameterization of $\mathbf{C}$ is, the appropriate Wigner $\mathscr{D}^1$ matrix can be written as

$$\mathscr{D}^1 = \mathbf{X}^\dagger \mathbf{C} \mathbf{X} \tag{49}$$

with

$$\mathbf{X} = \begin{pmatrix} -1/\sqrt{2} & 0 & 1/\sqrt{2} \\ -i/\sqrt{2} & 0 & -i/\sqrt{2} \\ 0 & 1 & 0 \end{pmatrix} \tag{50}$$

provided that $C_{\alpha\beta}$ is ordered $\alpha, \beta = x, y, z$ and the indices on $\mathscr{D}^1$ run $+1, 0, -1$ across each row and down each column.

The elements of the general matrix $\mathscr{D}^J$ can then be obtained (see Appendix 2 of Ref. 17) by repeated vector-coupling of the elements of $\mathscr{D}^1$.

It is thus perfectly possible to get expressions for the angular momentum eigenfunctions directly in terms of the elements of **C**. (For details, see Section 6.19 of Ref. 16.) This possibility is extensively exploited in Louck's derivation[18] of the Eckart Hamiltonian.

The result (49) is useful when expressing the dipole operator in body-fixed coordinates because it can be seen from (23) and (25) that

$$\mathbf{d}(\mathbf{t}) = e \sum_{j=1}^{N-1} \tilde{Z}_j \mathbf{C} \mathbf{z}_j \tag{51}$$

so that

$$\mathbf{X}^\dagger \mathbf{d}(\mathbf{t}) = e \sum_{j=1}^{N-1} \tilde{Z}_j \mathscr{D}^1 \mathbf{X}^\dagger \mathbf{z}_j \tag{52}$$

It is not the practice to work with the form $\mathbf{X}^\dagger \mathbf{d}$ but rather with the so-called "spherical" form, $\mathbf{X}^T \mathbf{d}$, so that, for example,

$$\mathbf{X}^T \mathbf{z} = \begin{pmatrix} -(z_x + i z_y)/\sqrt{2} \\ z_z \\ (z_x - i z_y)/\sqrt{2} \end{pmatrix} \tag{53}$$

Therefore, taking the complex conjugate of both sides of (52) gives

$$\mathbf{d}(\mathbf{t}) = \mathscr{D}^{1*} \mathbf{d}(\mathbf{q}) \tag{54}$$

where the assumption is that the elements of **d** are given in spherical form and labeled (d_{+1}, d_0, d_{-1}) and where

$$\mathbf{d}(\mathbf{q}) = e \sum_{j=1}^{N-1} \tilde{Z}_j \mathbf{X}^T \mathbf{z}_j \tag{55}$$

The form (54) of the dipole operator enables matrix elements to be calculated over the angular functions by use of straightforward angular momentum algebra. Thus, using the standard expression for a Gaunt coefficient

$$\langle J'M'k' | \mathscr{D}^{1*}_{mp} | JMk \rangle$$
$$= (-1)^{M'+k}[(2J'+1)(2J+1)]^{1/2}\begin{pmatrix} J' & 1 & J \\ -M' & m & M \end{pmatrix}\begin{pmatrix} J' & 1 & J \\ k' & p & -k \end{pmatrix} \tag{56}$$

where the $3-j$ symbols are as defined in Ref. 15.

The transformation of the space-fixed kinetic energy operator into body-fixed form is a rather lengthy and tedious business to give in detail. However, the final result can be stated reasonably straightforwardly, and because the derivation itself is quite mechanical, involving only letting (36) operate on itself and summing over i and j as in (11), there is no need to go into details. The resulting body-fixed operator is

$$\hat{K}(\boldsymbol{\phi}, \mathbf{q}) = \frac{1}{2}\left(\sum_{\alpha\beta} M_{\alpha\beta}\hat{L}_\alpha\hat{L}_\beta + \hbar\sum_\alpha \lambda_\alpha \hat{L}_\alpha\right)$$
$$-\frac{\hbar^2}{2}\left(\sum_{k,l=1}^{3N-6} G_{kl}\frac{\partial^2}{\partial q_k\,\partial q_l} + \sum_{k=1}^{3N-6}\tau_k\frac{\partial}{\partial q_k}\right) \tag{57}$$

whereas the potential energy operator is built just as in (11) by means of (13) but now with $\mathbf{z}_k$ replacing $\mathbf{t}_k$ in (13) to give $f_{ij}(\mathbf{q})$.

The matrix $\mathbf{M}$ is a generalized inverse inertia tensor defined as

$$\mathbf{M} = \sum_{i,j=1}^{N-1} \mu_{ij}^{-1}\boldsymbol{\Omega}^{iT}\boldsymbol{\Omega}^j \tag{58}$$

while $\mathbf{G}$ is given by

$$\mathbf{G} = \sum_{i,j=1}^{N-1} \mu_{ij}^{-1}\mathbf{Q}^{iT}\mathbf{Q}^j \tag{59}$$

In the term linear in the angular momentum

$$\lambda^\alpha = \frac{1}{i}\left(v_\alpha + 2\sum_{k=1}^{3N-6} W_{k\alpha}\frac{\partial}{\partial q_k}\right) \tag{60}$$

with

$$\mathbf{W} = \sum_{i,j=1}^{N-1} \mu_{ij}^{-1} \mathbf{Q}^{iT} \mathbf{\Omega}^{j} \tag{61}$$

and

$$v_{\alpha} = \sum_{i,j=1}^{N-1} \mu_{ij}^{-1} \left\{ \sum_{\beta} \left[(\mathbf{\Omega}^{iT} \mathbf{M}^{\beta} \mathbf{\Omega}^{j})_{\beta\alpha} + \sum_{l=1}^{3N-6} Q^{i}_{\beta l} \frac{\partial}{\partial q_l} \Omega^{j}_{\beta\alpha} \right] \right\} \tag{62}$$

Equation (60) is associated with the Coriolis coupling and so no coordinate system can be found in which it will vanish.

In the term linear in the derivatives of the q_k

$$\tau_{k} = \sum_{i,j=1}^{N-1} \mu_{ij}^{-1} \left\{ \sum_{\beta} \left[(\mathbf{\Omega}^{iT} \mathbf{M}^{\beta} \mathbf{Q}^{j})_{\beta k} + \sum_{l=1}^{3N-6} Q^{i}_{\beta l} \frac{\partial}{\partial q_l} Q^{j}_{\beta k} \right] \right\} \tag{63}$$

It is possible to choose a coordinate system in which this term vanishes.

It is clear from what has gone before—and indeed from general group theoretical arguments—that the eigenfunctions $\Psi(\mathbf{t})$ from (14) can be written as eigenfunctions of the body-fixed Hamiltonian in the form

$$\Psi(\mathbf{t}) \rightarrow \Psi^{J,M}(\boldsymbol{\phi}, \mathbf{q}) = \sum_{k=-J}^{+J} \Phi^{J}_{k}(\mathbf{q}) \, |JMk\rangle \tag{64}$$

where the internal coordinate function on the right-hand side cannot depend on M because, in the absence of a field, the energy of the system does not depend on M, as will be seen.

It is thus possible to eliminate the angular motion from the problem and to express an effective body-fixed Hamiltonian within any (J, M, k) rotational manifold, which depends only on the internal coordinates. Before showing how this is to be done it is appropriate to say something about the Jacobian to be used in performing the integrals. Clearly, the volume element for space-fixed integration is just $d\mathbf{t}_1 d\mathbf{t}_2, \ldots, d\mathbf{t}_{N-1}$ because the transformation from the laboratory-fixed to the space-fixed frame is linear and so $|\mathbf{V}^{-1}| = |\mathbf{V}^{-1}|$ is simply a constant that can be ignored. The transformation from the space-fixed coordinates to the body-fixed ones is nonlinear and so

$$d\mathbf{t}_1 \, d\mathbf{t}_2, \ldots, d\mathbf{t}_{N-1} = |\mathbf{J}|^{-1} \, d\phi_1 \, d\phi_2 \, d\phi_3 \, dq_1 \, dq_2, \ldots, dq_{3N-6}$$

or

$$dt = |\mathbf{J}|^{-1}\, d\boldsymbol{\phi}\, d\mathbf{q} \tag{65}$$

where $|\mathbf{J}|$ is the determinant of the matrix constructed from (34) and (35). Of course, $|\mathbf{J}^{-1}|$ could have been used equally well where the determinant would have been constructed from (38) and (39), but in practice it is necessary to construct the $\boldsymbol{\Omega}^i$ and $\mathbf{Q}^i$ to get explicit forms for the operators, so it is easier to use these.

In many cases it is actually more convenient to construct the determinant of the metric derived from (34) and (35), which will be equal to $|\mathbf{J}|^{-2}$, to within a constant factor. The determinant is (see Appendix 1 at the end of this chapter)

$$|\mathbf{D}^{-1}|^2 \begin{vmatrix} \mathbf{M} & \mathbf{W}^T \\ \mathbf{W} & \mathbf{G} \end{vmatrix}^{-1} \tag{66}$$

where the matrices in the partitions are given by (58), (59), and (61). Simple algebra then shows that (again to within a constant factor)

$$|\mathbf{J}|^{-1} = |\mathbf{D}|^{-1}|\mathbf{M}|^{-1/2}|\mathbf{G} - \mathbf{W}\mathbf{M}^{-1}\mathbf{W}^T|^{-1/2} \tag{67}$$

It is the factor $|\mathbf{D}|^{-1}$ that is the angular part of the Jacobian, and it is easy to show that in the standard parameterization of Ref. 15, $|\mathbf{D}|^{-1} = \sin\phi_2$ as required for the usual interpretation of the matrix element. The remaining terms in (67) are functions of the q_k alone. It is perhaps worthwhile noting that in the Podolsky approach to the construction of the body-fixed Hamiltonian, which is the approach used by Watson[(7)] (and indeed many others) it is the form of the Jacobian arising from (67) that is used.

The choice is sometimes made to incorporate the internal coordinate part of the Jacobian (or some of it) into the definition of the Hamiltonian. This is a fairly familiar process when working in spherical polars, for example, where the radial volume element r^2dr can be reduced to dr by writing the trial wavefunction $\psi(r)$ as $r^{-1}P(r)$ and modifying the Hamiltonian to refer to $P(r)$. This modification changes the derivative terms in the operator by $\partial/\partial r \rightarrow (\partial/\partial r - 1/r)$ and so on, but it alters none of the multiplicative or $\partial/\partial\theta$ terms. The resulting Hamiltonian is often said to be in manifestly Hermitian form. Particular examples of this kind of construction can be found in Watson[(7)] and Louck,[(18)] whereas a general account is given in Section 35 of Kemble.[(19)] This process is often extremely useful in practice with specific coordinate choices; however, it does not simplify matters at the level of formal exposition, so it will not be considered further here.

To remove the rotational motion it is convenient to write $\hat{K}(\boldsymbol{\phi}, \mathbf{q})$ given by (57) in two parts corresponding to the bracketing in that equation

$$\hat{K}_R(\boldsymbol{\phi}, \mathbf{q}) + \hat{K}_I(\mathbf{q}) \tag{68}$$

and to write the potential as $V(\mathbf{q})$. The matrix elements with respect to the angular functions of the operators that depend only on the q_k are quite trivial. Thus,

$$\langle J'M'k' | \hat{K}_I + V | JMk \rangle = \delta_{J'J}\, \delta_{M'M}\, \delta_{k'k} (\hat{K}_I + V) \tag{69}$$

In what follows, explicit allowance for the diagonal requirement on J and M will be assumed, and the indices will be suppressed. Similarly, that the integration is over $\boldsymbol{\phi}$ will only be implied.

To address the first term in (68) is considerably more complicated and is best done by reexpressing the components of $\hat{\mathbf{L}}$ in terms of $\hat{L}_\pm(\boldsymbol{\phi})$ and $\hat{L}_z(\boldsymbol{\phi})$ and using (45) and (46). When this is done,

$$\begin{aligned}
\langle JMk' | \hat{K}_R | JMk \rangle = {} & \frac{\hbar^2}{4} (b_{+2} C^+_{Jk+1} C^+_{Jk}\, \delta_{k'k+2} + b_{-2} C^-_{Jk-1} C^-_{Jk}\, \delta_{k'k-2}) \\
& + \frac{\hbar^2}{4} \{ C^+_{Jk} [b_{+1}(2k+1) + \lambda_+]\, \delta_{k'k+1} \\
& + C^-_{Jk} [b_{-1}(2k-1) + \lambda_-]\, \delta_{k'k-1} \} \\
& + \frac{\hbar^2}{2} \{ [J(J+1) - k^2] b + b_0 k^2 + \lambda_0 k \}\, \delta_{k'k}
\end{aligned} \tag{70}$$

In this expression

$$\begin{aligned}
b_{\pm 2} &= \frac{M_{xx} - M_{yy}}{2} \pm \frac{M_{xy}}{i} \\
b_{\pm 1} &= M_{xz} \pm \frac{M_{yz}}{i} \\
b &= \frac{M_{xx} + M_{yy}}{2}, \qquad b_0 = M_{zz}
\end{aligned} \tag{71}$$

and in terms of the λ_a in (60), λ_0 is λ_z and the $\lambda_\pm$ are

$$\lambda_\pm = \left(\lambda_x \pm \frac{\lambda_y}{i}\right) \tag{72}$$

Thus, within any rotational manifold it is the eigensolutions of the effective Hamiltonian given by (69) and (70) that are invariant to orthogonal transformations.

It is similarly possible to derive an expression for the effective dipole moment operator, but here *two* rotation manifolds are connected because of the nature of the Gaunt coefficient in (56). Using an explicit notation to deal with this, we obtain

$$\begin{aligned}
&\langle \Psi'(\mathbf{t}) | d_m(\mathbf{t}) | \Psi(t) \rangle \\
&= [(2J'+1)(2J+1)]^{1/2}(-1)^{M'} \begin{pmatrix} J' & 1 & J \\ -M' & m & M \end{pmatrix} \\
&\times \sum_{p=-1}^{+1} \sum_{k'=-J'}^{+J'} \sum_{k=-J}^{+J} (-1)^k \begin{pmatrix} J' & 1 & J \\ k' & p & -k \end{pmatrix} \langle \Phi_{k'}^{\prime J'} | d_p(\mathbf{q}) | \Phi_k^J \rangle
\end{aligned} \tag{73}$$

where the spherical form (53) has been used for the components of the dipole. For the equivalent expression using the more conventional angular momentum eigenfunctions, see Section 7.10 of Ref. 16. The expressions differ only by a phase factor and this is of no account in the calculations of line strengths.

It should be noted that it is not absolutely necessary to perform the passage to body-fixed coordinates in two stages and that one can pass directly to the body-fixed frame if desired. It is also the case that alternative forms to that given in (57) for the kinetic energy operator can be found and are sometimes useful. Brief accounts of these possibilities are given in Appendix 1 at the end of this chapter.

The underlying group structure implicit in the body-fixing as exhibited so far is $SO(3)$ rather than the full orthogonal group $O(3)$. The necessary extension involves an explicit consideration of parity and is somewhat involved. It has rather limited consequences, excepting at an epistemological level, and is confined to Appendix 2.

It was remarked in explaining (65) that the transformation from the space-fixed to the body-fixed coordinates is nonlinear and that it can be shown this is a topological consequence of any transformation that allows rotational motion to be separated. Furthermore, it can be shown that there is always some configuration of the particles that causes the Jacobian to vanish. Clearly, where the Jacobian vanishes, the transformation is undefined. These and related matters are discussed in more detail in Sutcliffe.[20] What it means in the present context is that not all possible $\Phi_k^J(\mathbf{q})$ are valid trial functions in (64)—only those that are strongly vanishing where the Jacobian vanishes. If an illegitimate trial function is used in practice, divergent expectation values of the Hamiltonian result.

It should be stressed that the origin of these divergences is not physical; they arise simply as a consequence of coordinate choice. However, a particular choice can obviously preclude the description of a possible physical state of a system. Thus, suppose that a triatomic is described in the Eckart approach with the equilibrium geometry specified as bent. In this case the Jacobian vanishes when the internal coordinates correspond to a linear geometry. The problem in this formulation then becomes ill-conditioned for states with large-amplitude bending motions. Of course, such large-amplitude bending states are physically perfectly reasonable, it is just that they cannot be described in this formulation. A detailed discussion of these matters and an account of some proposals for coping with them in practice for triatomics is given in Refs. 14, 21, and 22. The underlying problem is well known and has long been studied.

From a physical point of view, therefore, it is hopeless to attempt to describe all possible bound states $\Psi(\mathbf{t})$ in the form of (64) by means of a *single* choice of internal and angular coordinates. In fact, rather detailed topological arguments (which are summarized in Ref. 20) show that this physical sense is mathematically sound and that more than one set of body-fixed coordinates is always required to describe the space-fixed set over its whole range. However, in any particular case, a set of body-fixed coordinates can be found in which (64) is a valid expression.

It would be wrong to end this section without making it quite clear that although the transformation processes outlined above are formally impeccable, their implementation in practice can be extremely difficult or even impossible. The nature of the difficulties will be apparent to those familiar with the work of Watson[7] and Louck,[18] and an example of where things get impossible can be found in Carter and Handy.[23] It has been suggested by Handy[24] that computer-aided algebra systems might help to mitigate the difficulties, but it is well known that it is not always possible to invert a nonlinear transformation explicitly, so it is likely that there will always remain some desirable coordinate choices that are impossible in practice.

It is now necessary to see how far it is possible to proceed in separating electronic and nuclear motions to yield an electronic problem that can be made to correspond to the usual clamped-nucleus Hamiltonian and thus to yield a potential function that is invariant to translations and to orthogonal transformations of the laboratory-fixed coordinates.

3. Separating Electronic and Nuclear Motions

It should be clear that the identification of electronic and nuclear variables among space-fixed variables is purely conventional. No matter how this identification is made, the energy spectrum of the space-fixed Hamiltonian remains the same. Thus, the choice can be made simply on the grounds of convenience and utility while recognizing that there can be disagreement about what is a reasonable choice. Therefore, what follows in this section cannot, by definition as it were, be regarded as definitive. It is believed, however, to be accurate.

Because the aim is to explicate the standard clamped-nucleus electronic Hamiltonian, it seems reasonable to choose the same number of space-fixed electronic variables as there are electrons in the problem. Let that number be L. If there are H nuclei in the problem ($N = H + L$), then there can only be $H - 1$ space-fixed nuclear variables. It is assumed that H is at least 2. Let the set of laboratory fixed coordinates be split into an electronic set, $\mathbf{x}_1^e$, $i = 1, 2, \ldots, L$ and a nuclear set $\mathbf{x}_i^n$, $i = 1, 2, \ldots, H$. The electronic charge numbers will be taken explicitly as -1, and the electronic mass will be written as m. The charge numbers Z_i and the masses m_i will be taken as referring to the nuclei only.

It seems reasonable to require that the space-fixed nuclear coordinates be expressible entirely in terms of the laboratory-fixed nuclear coordinates. Thus, analogously to (3),

$$(\mathbf{t}^n\mathbf{X}) = \mathbf{x}^n\mathbf{V}^n \tag{74}$$

Here $\mathbf{t}^n$ is a $3 \times H - 1$ matrix, and $\mathbf{X}$ is a 3×1 matrix. $\mathbf{V}^n$ is an $H \times H$ matrix whose last column is special, with elements

$$V_{iH}^n = M^{-1}m_i, \qquad M = \sum_{i=1}^{H} m_i \tag{75}$$

so that $\mathbf{X}$ is the center-of-nuclear-mass coordinate. The elements in each of the first $H - 1$ columns of $\mathbf{V}^n$ sum to zero, precisely as in (6), to ensure translation invariance.

If the $\mathbf{t}^n$ are independent, then

$$\mathbf{x}^n = (\mathbf{t}^n\mathbf{X})(\mathbf{V}^n)^{-1} \tag{76}$$

just as in (7) and, just as in (8), the bottom row of $(\mathbf{V}^n)^{-1}$ is special with elements

$$[(\mathbf{V}^n)^{-1}]_{Hi} = 1, \qquad i = 1, 2, \ldots, H \tag{77}$$

whereas like (9) the inverse requirements on the remaining rows gives

$$\sum_{i=1}^{H} [(\mathbf{V}^n)^{-1}]_{ji} m_i = 0, \qquad j = 1, 2, \ldots, H-1 \tag{78}$$

The space-fixed electronic coordinates will have to involve the laboratory-fixed nuclear coordinates so that (74) may be generalized as:

$$(\mathbf{t}^e \mathbf{t}^n \mathbf{X}) = (\mathbf{x}^e \mathbf{x}^n) \begin{pmatrix} \mathbf{V}^e & \mathbf{0} \\ \mathbf{V}^{ne} & \mathbf{V}^n \end{pmatrix} \tag{79}$$

where $\mathbf{t}^e$ is a $3 \times L$ matrix. It is not possible to choose $\mathbf{V}^{ne}$ to be a null matrix and to satisfy simultaneously the translational invariance requirements as shown in (6) while requiring the matrix (79) to be nonsingular. Given that the inverse of (79) exists, however, then (76) may be generalized as

$$(\mathbf{x}^e \mathbf{x}^n) = (\mathbf{t}^e \mathbf{t}^n \mathbf{X}) \begin{pmatrix} (\mathbf{V}^e)^{-1} & \mathbf{0} \\ \mathbf{B} & (\mathbf{V}^n)^{-1} \end{pmatrix} \tag{80}$$

where

$$\mathbf{B} = -(\mathbf{V}^n)^{-1} \mathbf{V}^{ne} (\mathbf{V}^e)^{-1} \tag{81}$$

The bottom row of $\mathbf{B}$ is special in the same way as is the bottom row of $(\mathbf{V}^n)^{-1}$ and consists of the elements

$$B_{Hi} = 1, \qquad i = 1, 2, \ldots, H \tag{82}$$

Using (80) and the definitions of $\mathbf{X}$ and $\mathbf{X}_T$ it follows that

$$\mathbf{X}_T = \mathbf{X} + \sum_{i=1}^{L} s_i^e \mathbf{t}_i^e + \sum_{i=1}^{H-1} s_i^n \mathbf{t}_i^n \tag{83}$$

where

$$s_i^e = M_T^{-1} m \sum_{j=1}^{L} [(V^e)^{-1}]_{ij}, \qquad s_i^n = M_T^{-1} m \sum_{j=1}^{L} B_{ij} \tag{84}$$

so that using (83) $\mathbf{X}$ can be eliminated in favor of $\mathbf{X}_T$ whenever necessary.

In fact, it is obvious from the definition (79) that a change from $\mathbf{X}$ to $\mathbf{X}_T$ has no effect on the expression for $\mathbf{t}^e$ and $\mathbf{t}^n$, but from (80) it is seen that such a change does affect the expression for the inverse. Using (80) and (83)

$$\mathbf{x}_i^e = \mathbf{X}_T + \sum_{j=1}^{L} \mathbf{t}_j^e[(\mathbf{V}^e)_{ji}^{-1} - s_j^e] + \sum_{j=1}^{H-1} \mathbf{t}_j^n(B_{ji} - s_j^n) \tag{85}$$

and

$$\mathbf{x}_i^n = \mathbf{X}_T - \sum_{j=1}^{L} \mathbf{t}_j^e s_j^e + \sum_{j=1}^{H-1} \mathbf{t}_j^n[(\mathbf{V}^n)_{ji}^{-1} - s_j^n] \tag{86}$$

If equation (85) is substituted into $\mathbf{x}_{ij}^e = |\mathbf{x}_j^e - \mathbf{x}_i^e|$, then the resulting form is not generally expressible in terms of the $\mathbf{t}_i^e$ alone. Furthermore, the form is not invariant under permutation of identical nuclei and neither does it change in the ordinary way under the permutation of electrons. These features are obviously undesirable for they are not reflected in the standard expression for the clamped-nucleus Hamiltonian. So, it would be wise to restrict the form of $\mathbf{V}^{ne}$ and of $\mathbf{V}^e$.

To construct a transformation such that the space-fixed electronic coordinates are invariant under any permutation of identical nuclei so that under such a permutation the $\mathbf{t}_i^e$ are unaffected and also such that they change in the usual manner under permutation of laboratory-fixed electronic variables, consider the general permutation of identical particles. This can be written as

$$\mathscr{P}(\mathbf{x}^e\mathbf{x}^n) = (\mathbf{x}^e\mathbf{x}^n)\begin{pmatrix} \mathbf{P}^e & \mathbf{0} \\ \mathbf{0} & \mathbf{P}^n \end{pmatrix} \tag{87}$$

where $\mathbf{P}^e$ and $\mathbf{P}^n$ are standard permutation matrices.

Using (79) and (80), it follows that

$$\mathscr{P}(\mathbf{t}^e\mathbf{t}^n\mathbf{X}) = (\mathbf{t}^e\mathbf{t}^n\mathbf{X})\begin{pmatrix} (\mathbf{V}^e)^{-1} & \mathbf{0} \\ \mathbf{B} & (\mathbf{V}^n)^{-1} \end{pmatrix}\begin{pmatrix} \mathbf{P}^e & \mathbf{0} \\ \mathbf{0} & \mathbf{P}^n \end{pmatrix}\begin{pmatrix} \mathbf{V}^e & \mathbf{0} \\ \mathbf{V}^{ne} & \mathbf{V}^n \end{pmatrix} \tag{88}$$

To achieve the required invariance, the matrix on the right-hand side of (88) must be block diagonal, and this occurs only if

$$\mathbf{B}\mathbf{P}^e\mathbf{V}^e + (\mathbf{V}^n)^{-1}\mathbf{P}^n\mathbf{V}^{ne} = \mathbf{0}_{H,L} \tag{89}$$

The most general way in which this can be achieved is to require the following relations to hold

$$\mathbf{P}^e\mathbf{V}^e = \mathbf{V}^e\mathbf{P}^e \tag{90}$$

$$\mathbf{P}^n\mathbf{V}^{ne} = \mathbf{V}^{ne} \tag{91}$$

$$\mathbf{V}^{ne}\mathbf{P}^e = \mathbf{V}^{ne} \tag{92}$$

If these relations hold, then $\mathbf{P}^e$ can be taken as a common factor to the right and, using (81) for $\mathbf{B}$, $(\mathbf{V}^n)^{-1}$ may be taken as a common factor to the left in (89). The factored matrix then vanishes identically.

The physical content of the requirement in (91) is simply that every member of a set of identical nuclei must enter into the definition of $\mathbf{t}^e$ in the same way. The physical content of the requirement in (92) is that any electronic variable in the problem should have exactly the same relationship to the nuclear variables as does any other electronic variable. Thus, (92) is satisfied by requiring all the columns of $\mathbf{V}^{ne}$ to be identical, and from now on a typical column will be denoted $\mathbf{v}$.

The most general form for $\mathbf{V}^e$ that satisfies (90) is

$$(\mathbf{V}^e)_{ij} = \delta_{ij} + a \tag{93}$$

where a is a constant. The inverse has elements

$$[(\mathbf{V}^e)^{-1}]_{ij} = \delta_{ij} - \frac{a}{(1 + La)} \tag{94}$$

This shows that a can take any value (including 0) except $-1/L$. Using these results in (81) it follows that the columns of $\mathbf{B}$ are identical with each other and the typical column will now be written as $\mathbf{b}$. Using (85) with these restrictions it follows that $\mathbf{x}_{ij}^e$ becomes $\mathbf{t}_{ij}^e$, as required. Using the results in (84) gives

$$s_i^e = \frac{m}{M_T(1 + La)}, \qquad s_i^n = M_T^{-1}Lmb_i, \qquad \mathbf{b} = -\frac{(\mathbf{V}^n)^{-1}\mathbf{v}}{1 + La} \tag{95}$$

If equations (90)–(92) are satisfied, then it is easy to show that if in (88), X is replaced by X_T, the equation generalizes to

$$\mathscr{P}(\mathbf{t}^e\mathbf{t}^n\mathbf{X}_T) = (\mathbf{t}^e\mathbf{t}^n\mathbf{X}_T)\begin{pmatrix} \mathbf{P}^e & \mathbf{0} & \mathbf{0} \\ \mathbf{0} & \mathbf{H} & \mathbf{0} \\ \mathbf{0} & \mathbf{0} & 1 \end{pmatrix} \tag{96}$$

where

$$(\mathbf{H})_{ij} = [(\mathbf{V}^n)^{-1}\mathbf{P}^n\mathbf{V}^n]_{ij}, \qquad i, j = 1, 2, \ldots, H-1 \tag{97}$$

The $(H-1) \times (H-1)$ matrix $\mathbf{H}$ is not generally in standard permutational form; neither is it orthogonal even though it has determinant ± 1 according to the sign of $|\mathbf{P}^n|$. It will be assumed that a coordinate system has been chosen according to (79) and (80) in which the conditions (90)–(92) are satisfied.

The above discussion gives an account of a reasonable and reasonably convenient way of partitioning $\mathbf{V}$ and hence its inverse, if the division of the problem into electronic and nuclear parts is to be recognized and made explicit. The form of the space-fixed operators remains unchanged from that given in the previous section, but the partition made here does enable a more specific structure to be given to them with parts attributable to the types of particle.

Thus, the derivative operator (10) can now be distinguished as consisting of two parts

$$\frac{\partial}{\partial \mathbf{x}_i^e} = mM_T^{-1}\frac{\partial}{\partial \mathbf{X}_T} + \sum_{j=1}^{L}(\delta_{ij} + a)\frac{\partial}{\partial \mathbf{t}_j^e} \tag{98}$$

$$\frac{\partial}{\partial \mathbf{x}_i^n} = m_i M_T^{-1}\frac{\partial}{\partial \mathbf{X}_T} + v_i \sum_{j=1}^{L}\frac{\partial}{\partial \mathbf{t}_j^e} + \sum_{j=1}^{H-1} V_{ij}^n \frac{\partial}{\partial \mathbf{t}_j^n} \tag{99}$$

The space-fixed Hamiltonian arising from the last two terms in (11) expands into three parts:

$$\hat{H}(\mathbf{t}) \rightarrow \hat{H}^e(\mathbf{t}^e) + \hat{H}^n(\mathbf{t}^n) + \hat{H}^{en}(\mathbf{t}^n, \mathbf{t}^e) \tag{100}$$

Here

$$\hat{H}^e(\mathbf{t}^e) = -\frac{\hbar^2}{2\mu}\sum_{i=1}^{L}\nabla^2(\mathbf{t}_i^e) - \frac{\hbar^2}{2\mu'}\sum_{ij=1}^{L}{}' \nabla(\mathbf{t}_i^e)\cdot\nabla(\mathbf{t}_j^e) + \frac{e^2}{8\pi\varepsilon_0}\sum_{ij=1}^{L}{}' \frac{1}{|\mathbf{t}_j^e - \mathbf{t}_i^e|} \tag{101}$$

with

$$\frac{1}{\mu} = \frac{1}{m} + \frac{1}{\mu'} \tag{102}$$

$$\frac{1}{\mu'} = a(2 + La)/m + \sum_{k=1}^{H} v_e^{+2}/m_k \tag{103}$$

and

$$\hat{H}^n(\mathbf{t}^n) = -\frac{\hbar^2}{2}\sum_{ij=1}^{H-1} \mu_{ij}^{-1}\boldsymbol{\nabla}(\mathbf{t}_i^n)\cdot\boldsymbol{\nabla}(\mathbf{t}_j^n) + \frac{e^2}{8\pi\varepsilon_0}\sum_{ij=1}^{H}{}' \frac{Z_iZ_j}{f_{ij}(\mathbf{t}^n)} \tag{104}$$

where μ_{ij}^{-1} is defined just as in (12) but in terms of the nuclear masses only and using $\mathbf{V}^n$. Similarly, $f_{ij}(\mathbf{t}^n)$ is defined just as in (13) but using the $\mathbf{t}_i^n$ only and $(\mathbf{V}^n)^{-1}$.

Finally,

$$\hat{H}^{en}(\mathbf{t}^n, \mathbf{t}^e) = -\hbar^2 \sum_{i=1}^{H-1} \mu_i^{-1}\boldsymbol{\nabla}(\mathbf{t}_i^n)\cdot\sum_{j=1}^{L}\boldsymbol{\nabla}(\mathbf{t}_j^e) - \frac{e^2}{4\pi\varepsilon_0}\sum_{i=1}^{H-1}\sum_{j=1}^{L}\frac{Z}{f'_{ij}(\mathbf{t}^n, \mathbf{t}^e)} \tag{105}$$

with

$$\mu_i^{-1} = \sum_{k=1}^{H} m_k^{-1} v_k V_{ki}^n \tag{106}$$

while f'_{ij} is the electron–nucleus distance and so it is the modulus

$$|\mathbf{x}_i^n - \mathbf{x}_j^e| = \left|\sum_{k=1}^{H-1} \mathbf{t}_k^n[(\mathbf{V}^n)_{ki}^{-1} - b_k] + \frac{a}{1 + La}\sum_{k=1}^{L}\mathbf{t}_k^e - \mathbf{t}_j^e\right| \tag{107}$$

The space-fixed angular momentum operator [the second term in (21)] can be written as

$$\hat{\mathbf{L}}(\mathbf{t}^n, \mathbf{t}^e) = \frac{\hbar}{i}\sum_{i=1}^{H-1}\hat{\mathbf{t}}_i^n\frac{\partial}{\partial\mathbf{t}_i^n} + \frac{\hbar}{i}\sum_{i=1}^{L}\hat{\mathbf{t}}_i^e\frac{\partial}{\partial\mathbf{t}_i^e} \tag{108}$$

The space-fixed dipole operator arising from the first term in (23) can similarly be written in two parts corresponding to the nuclear–electronic division as

$$\mathbf{d}(\mathbf{t}^n, \mathbf{t}^e) = e \sum_{i=1}^{H-1} \tilde{Z}_i \mathbf{t}_i^n - e\tilde{Z} \sum_{i=1}^{L} \mathbf{t}_i^e \tag{109}$$

where, using (24) with (85), (86), and (95),

$$\tilde{Z}_i = \sum_{j=1}^{H} [(\mathbf{V}^n)^{-1}]_{ij} Z_j - \frac{L(M_T + mZ_T)b_i}{M^T} \tag{110}$$

$$\tilde{Z} = \frac{M_T + mZ_T}{M_T(1 + La)} \tag{111}$$

The expressions so far obtained are reasonably general, if somewhat cumbersome. However, were it to be thought desirable to attempt to make ab initio calculations directly with the space-fixed form or to do body-fixed calculations similarly, then the rather awkward form of the Hamiltonian could have computational advantages. It is nevertheless possible to simplify it somewhat, and if one is thinking of an approach to the problem through the standard potential energy surface, then the simplification is actually advantageous. A more in-depth discussion of this will be given later.

Let the elements of $\mathbf{v}$ be chosen as

$$v_k = -\alpha M^{-1} m_k, \qquad \alpha = (1 + La) \tag{112}$$

where the choice for α is determined by the translational invariance requirements as explained in connection with (79). Thus, the $\mathbf{t}_i^e$ are the electronic coordinates referred to the center-of-nucleus-mass scaled by α. This choice also satisfies the condition expressed in equation (91), and from (78) and (95) $\mathbf{b} = \mathbf{0}$. Because of this, s_i^n vanishes and the electron–nucleus attraction term (107) and the effective nuclear charge term (110) are both somewhat simplified by the loss of the parts that depend on the elements of $\mathbf{b}$. Substituting (112) for v_k into (106), it follows at once that μ_i^{-1} vanishes so that the electron–nucleus coupling term in the kinetic energy part of (105) vanishes.

Assuming the choice in (112) to have been made, then a value for α or, equivalently, a value for a remain to be chosen. There are two choices that are of immediate interest; the first arises from an attempt to simplify equation (101) and the second from an attempt to simplify equation (107). For the first, consider equation (103) with (112) for v_k:

$$\mu'^{-1} = m^{-1}a(2 + La) + \alpha^2 M^{-1} \tag{113}$$

Writing

$$a = \frac{\alpha - 1}{L}, \qquad \alpha = \left(\frac{M}{M_T}\right)^{1/2} \tag{114}$$

it is seen that μ'^{-1} vanishes and μ becomes m.

This strategy for choosing a in relation to α is analogous to the Radau choice of heliocentric coordinates (see, e.g., Smith[(25)]), but here the center-of-nuclear-mass plays the part of the distinguished coordinate. It is clearly an extremely attractive choice for simplifying the electronic kinetic energy in (101), but it leaves intact the rather awkward electronic sum term in (107) and it does not much simplify the effective electronic charge in (111). Simplification of these two terms is best achieved by choosing $a = 0$ so that the electronic sum term in (107) vanishes and $\tilde{Z}$ becomes one if the system is neutral. If this choice is made it is necessary [see equation (112)] to choose $\alpha = 1$ so that $\mu' = M$ and off-diagonal terms remain in the kinetic energy in (101).

In what follows, although the center-of-nuclear mass choice will be made so that the kinetic energy term in (105) vanishes, no particular choice will be assumed either for α or a.

In moving to the body-fixed frame it would seem reasonable to require that the matrix $\mathbf{C}$ is specified entirely in terms of the space-fixed nuclear variables $\mathbf{t}^n$. This means that the case $H = 2$ must be excluded much as the case $N = 2$ had to be excluded in the previous section. The simplest system that can be considered, therefore, is a triatomic.*

The three relations required to define the body-frame arise therefore from three relations among the body-fixed nuclear Cartesian variables given by

$$\mathbf{z}_i^n = \mathbf{C}^T \mathbf{t}_i^n, \qquad i = 1, 2, \ldots, H - 1 \tag{115}$$

[See also equation (26).] There are therefore $3H - 6$ internal coordinates (which will be denoted q_k just as before) that are expressible as functions of the scalar products of the $\mathbf{t}_i^n$ in terms of which the $\mathbf{z}_i^n$ can be expressed completely. The remaining $3L$ internal coordinates are made up of the three Cartesian components of each of the L body-fixed electronic Cartesians

* In fact, even the triatomic is somewhat special because three points always define a plane, and the presence of a plane allows rather special mappings under the inversion operation. This is discussed in Appendix 2 at the end of this chapter.

given by:

$$\mathbf{z}_i = \mathbf{C}^T \mathbf{t}_i^e, \qquad i = 1, 2, \ldots, L \tag{116}$$

where there is no explicit superscript on the electronic body-fixed variables.

The Jacobian matrix elements in equation (34) are now in two parts:

$$\frac{\partial \phi_m}{\partial t_{\alpha i}^n} = (\mathbf{C}\boldsymbol{\Omega}^i\mathbf{D})_{\alpha m}, \qquad \frac{\partial \phi_m}{\partial t_{\alpha i}^e} = 0 \tag{117}$$

where the $\boldsymbol{\Omega}^i$ are functions of the q_k only.

The Jacobian matrix elements (35) develop into four parts:

$$\frac{\partial q_k}{\partial t_{\alpha i}^n} = (\mathbf{C}\mathbf{Q}^i)_{\alpha k}, \qquad \frac{\partial q_k}{\partial t_{\alpha i}^e} = 0 \tag{118}$$

where the $\mathbf{Q}^i$ are functions of the q_k only and

$$\frac{\partial z_{\gamma j}}{\partial t_{\alpha i}^n} = (\mathbf{C}\boldsymbol{\Omega}^i\hat{\mathbf{z}}_j)_{\alpha\gamma}, \qquad \frac{\partial z_{\gamma j}}{\partial t_{\alpha i}^e} = \delta_{ij} C_{\alpha\gamma} \tag{119}$$

It follows therefore that (36) becomes an expression in two parts:

$$\frac{\partial}{\partial \mathbf{t}_i^n} = \mathbf{C}\left(\boldsymbol{\Omega}^i\mathbf{D}\frac{\partial}{\partial \boldsymbol{\phi}} + \mathbf{Q}^i\frac{\partial}{\partial \mathbf{q}} + \boldsymbol{\Omega}^i \sum_{j=1}^{L} \hat{\mathbf{z}}_j \frac{\partial}{\partial \mathbf{z}_j}\right) \tag{120}$$

and

$$\frac{\partial}{\partial \mathbf{t}_i^e} = \mathbf{C}\frac{\partial}{\partial \mathbf{z}_i} \tag{121}$$

There are similar developments in the expressions associated with the inverse transformation so that (38) becomes

$$\frac{\partial t_{\alpha i}^n}{\partial \phi_m} = (\mathbf{C}\hat{\mathbf{z}}_i^n\mathbf{D}^{-T})_{\alpha m}, \qquad \frac{\partial t_{\alpha i}^e}{\partial \phi_m} = (\mathbf{C}\hat{\mathbf{z}}_i\mathbf{D}^{-T})_{\alpha m} \tag{122}$$

while (39) goes into four parts

$$\frac{\partial t_{\alpha i}^n}{\partial q_k} = (\mathbf{C}\tilde{\mathbf{Q}}^i)_{\alpha k}, \qquad \frac{\partial t_{\alpha i}^e}{\partial q_k} = 0 \tag{123}$$

and

$$\frac{\partial t^n_{\alpha i}}{\partial z_{\gamma j}} = 0, \qquad \frac{\partial t^e_{\alpha i}}{\partial z_{\gamma j}} = \delta_{ij} C_{\alpha\gamma} \tag{124}$$

The relationship between the Jacobian and its inverse leads to exactly the same form of equations as are given in (40) and (41), but now i and j are to be interpreted as referring to space-fixed nuclear coordinates and the sums are to run to $H-1$ rather than to $N-1$.

The angular momentum is still carried entirely by the Euler angles as in (43) and (44), and the generalization of the dipole operator follows in an obvious way on replacing (55) with (109). The body-fixed Hamiltonian now arises by transforming the terms in (100). The change in $\hat{H}^e(\mathbf{t}^e)$ as given by (101) is trivial and is effected simply by replacing $\mathbf{t}^e_i$ by $\mathbf{z}_i$ everywhere it occurs. The result will be denoted as $\hat{H}(\mathbf{z})$, dropping the superscript as the provenance of the operator is clear from the variable name. Furthermore, because the first term in (105) vanishes by construction, it is sufficient to transform (107), and this too is done simply by replacing $\mathbf{t}^e_i$ by $\mathbf{z}_i$ and $\mathbf{t}^n_i$ by $\mathbf{z}^n_i(\mathbf{q})$ wherever they occur. The transformation to the body-fixed form of the kinetic energy operator from (104) is a little more involved, however, because of the presence in the last term in (120) of body-fixed electronic variables. The required form is a slight extension of (57), namely,

$$\begin{aligned}\hat{K}(\boldsymbol{\phi}, \mathbf{q}, \mathbf{z}) = &\frac{1}{2}\left\{\sum_{\alpha\beta} M_{\alpha\beta}\hat{L}_\alpha\hat{L}_\beta + \hbar\sum_\alpha [\lambda_\alpha + 2(\mathbf{M}\hat{\mathbf{l}})_\alpha]\hat{L}_\alpha\right\} \\ &-\frac{\hbar^2}{2}\left(\sum_{k,l=1}^{3H-6} G_{kl}\frac{\partial^2}{\partial q_k\,\partial q_l} + \sum_{k=1}^{3H-6}\tau_k\frac{\partial}{\partial q_k}\right) \\ &+\frac{\hbar^2}{2}\left(\sum_{\alpha\beta} M_{\alpha\beta}\hat{l}_\alpha\hat{l}_\beta + \sum_\alpha \lambda_\alpha\hat{l}_\alpha\right)\end{aligned} \tag{125}$$

where $\hat{\mathbf{l}}$ is a 3×1 column matrix of Cartesian components:

$$\hat{\mathbf{l}} = \frac{1}{i}\sum_{i=1}^{L} \hat{\mathbf{z}}_i \frac{\partial}{\partial \mathbf{z}_i} \tag{126}$$

The expressions for $\mathbf{M}$, $\mathbf{G}$, and $\mathbf{W}$ are of exactly the same form as given in (58), (59), and (61), as are the expressions for λ_α, ν_α, and τ_k of the same form as given in (60), (62), and (63). However [see discussion following equation (104)], the sums extend only to $H - 1$, and μ_{ij}^{-1} is defined using $\mathbf{V}^n$ and the nuclear masses only.

The last term in (125), which couples the body-fixed electronic and nuclear variables, is composed of operators that define the Coriolis coupling and so cannot be transformed away by some particular coordinate choice. Had the kinetic energy operator (105) not been transformed away by means of the center-of-nuclear-mass choice (112), then the coupling expression found at this stage would have been more extensive and would have involved mixed derivatives in the body-fixed electronic and nuclear variables. The nuclear repulsion term is found by expressing the $\mathbf{t}^n$ in terms of the q_k to yield $f_{ij}(\mathbf{q})$ as in the discussion of (104).

The overall structure of the body-fixed problem is essentially preserved under the identification of particle types so that the separation is much as before. Equation (68) can now be written as

$$\hat{K}_R(\boldsymbol{\phi}, \mathbf{q}, \mathbf{z}) + \hat{K}_I(\mathbf{q}, \mathbf{z}) \tag{127}$$

where the first term is just the term in brackets that begins (125), and the second term consists of the remaining terms from that equation together with the kinetic energy operator from $\hat{H}(\mathbf{z})$. It is useful to recognize a division of V in (69) by writing it as

$$V \rightarrow V(\mathbf{q}) - V(\mathbf{q}, \mathbf{z}) + V(\mathbf{z}) \tag{128}$$

The first term is the nuclear repulsion potential term from (105) realized as a function of the q_k; the second term is the electron–nucleus attraction potential term from (107) realized as a function of the q_k and the $\mathbf{z}_i$; and the last term is just the electronic repulsion term from (101) expressed in terms of the $\mathbf{z}_i$.

Similarly, it is useful to expand the notation of (64) somewhat so that

$$\Psi^{J,M}(\boldsymbol{\phi}, \mathbf{q}, \mathbf{z}) = \sum_{k=-J}^{+J} \Phi_k^J(\mathbf{q}, \mathbf{z}) \,|JMk\rangle \tag{129}$$

and using functions like this as the basis in removing the rotational motion, the process yields equations exactly like (69) and (70). But making explicit the electronic variables in the second term in (125) means that the λ_α term in (70) has to be slightly extended:

$$\lambda_\alpha \rightarrow \lambda_\alpha + 2(\mathbf{M}\hat{\mathbf{l}})_\alpha \tag{130}$$

The volume element for integration is a simple development of (65) namely:

$$dt = |\mathbf{J}|^{-1}\, d\boldsymbol{\phi}\, d\mathbf{q}\, d\mathbf{z} \tag{131}$$

where $|\mathbf{J}|$ is exactly as given by (67) but with the constituent matrices as functions of the nuclear variables alone [see discussion following equation (104)], and thus the volume element for $\mathbf{z}$ is in standard Cartesian form.

To effect the uncoupling of the nuclear ($\mathbf{q}$) part of the problem from the electronic ($\mathbf{z}$) part in the manner proposed by Born and Huang[10] is formally quite similar to the uncoupling of the rotation ($\boldsymbol{\phi}$) part from the internal ($\mathbf{q}$) part as undertaken above. The internal motion function from (129) is expressed in terms of a sum of products of the form

$$\boldsymbol{\Phi}^{J}_{kp}(\mathbf{q})\,\psi_p(\mathbf{q}, \mathbf{z}) \tag{132}$$

where p labels the electronic state, and the sum is over p. The function $\psi_p(\mathbf{q}, \mathbf{z})$ is assumed known, just as $|JMk\rangle$ is assumed known. Also, the effective nuclear motion Hamiltonian is obtained in terms of matrix elements of the effective internal motion Hamiltonian between the $\psi_p(\mathbf{q}, \mathbf{z})$ with respect to the variables $\mathbf{z}$, just as the effective internal motion Hamiltonian itself is expressed in terms of matrix elements of the full body-fixed Hamiltonian between the $|JMk\rangle$ with respect to the $\boldsymbol{\phi}$. The effective nuclear motion Hamiltonian then contains the electronic state labels p as parameters, in much the same way that the full effective Hamiltonian for internal motion contains the angular momentum labels k. Of course, the analogy between the two derivations is simply a formal one. There is no underlying symmetry structure in the effective nuclear problem, and neither is the sum over p of definite extent as is the sum over k.

In fact, Hunter[26] has shown (at least for the case $J = 0$) that the *exact* wave function can be written as a single product of this form. However, in Hunter's form, ψ is not determined as the solution of any sort of electronic problem but is rather obtained as a conditional probability amplitude by a process of integration and is to be associated with a marginal probability amplitude $\boldsymbol{\Phi}$ to constitute a complete probability amplitude. The work of Czub and Wolniewicz[27] would seem to indicate, however, that it would be very difficult to use this scheme to define, ab initio, a potential in terms of which nuclear motion functions could be calculated. Thus, unless the full function is known it does not seem possible to determine its parts factored in this way. For all practical purposes, then, we must use the full Born and Huang approach. In the original formulation it was stipulated that the known functions, $\psi_p(\mathbf{q}, \mathbf{z})$, were to be looked on as exact solutions of a problem like

$$[\hat{H}(\mathbf{z}) - V(\mathbf{q}, \mathbf{z})]\psi_p(\mathbf{q}, \mathbf{z}) = E_p(\mathbf{q})\Psi_p(\mathbf{q}, \mathbf{z})$$

that is:

$$\hat{H}^{\text{elec}}(\mathbf{q}, \mathbf{z})\psi_p(\mathbf{q}, \mathbf{z}) = E_p(\mathbf{q})\psi_p(\mathbf{q}, \mathbf{z}) \tag{133}$$

Because in this equation there are no terms that involve derivatives with respect to the q_k, there is no development with respect to $\mathbf{q}$ in $E_p(\mathbf{q})$ or $\psi_p(\mathbf{q}, \mathbf{z})$. Thus, the $\mathbf{q}$ act here simply as parameters that can be chosen at will.

In fact, it is not absolutely essential for what follows to require the ψ_p to be eigenfunctions of $\hat{H}^{\text{elec}}$. A reasonably concise and useful form can be obtained simply by requiring that

$$\int \psi_{p'}^*(\mathbf{q}, \mathbf{z})\psi_p(\mathbf{q}, \mathbf{z})\, d\mathbf{z} \equiv \langle \psi_{p'} | \psi_p \rangle_z = \delta_{p'p} \tag{134}$$

and, using the above abbreviation to denote integration over all $\mathbf{z}$ only,

$$\langle \psi_{p'} | \hat{H}^{\text{elec}} | \psi_p \rangle_{\mathbf{z}} = \delta_{p'p} E_p(\mathbf{q}) \tag{135}$$

The requirements in equations (134) and (135) can be met in a simple and practical way by requiring the ψ_p to be solutions of a linear variation problem with matrix elements determined by integration over the $\mathbf{z}$ alone, for each and every value assigned to $\mathbf{q}$.

The effective nuclear motion Hamiltonian that depends only on the $\mathbf{q}$ is expressed in terms of matrix elements of the Hamiltonian (127) between pairs of functions like (129) but with internal coordinate parts like (132) integrated over the $\mathbf{z}$ as well as the angular factors. Doing this yields equations rather like (61) and (62) but, as explained above, coupled not only between different rotational states, labeled by k, but also between different electronic states, labeled by p. In deriving them it should be remembered that the product rule must be used when considering the effect of derivative operators with respect to the q_k because *both* terms in the product (132) depend on the $\mathbf{q}$ variables.

The term analogous to (69) becomes

$$\begin{aligned}\langle JMk'p' | \hat{K}_I - V(\mathbf{q}, \mathbf{z}) + V(\mathbf{q}) | JMkp \rangle_z \\ = \delta_{p'p}\, \delta_{k'k}[\hat{K}_H + E_p(\mathbf{q}) + V(\mathbf{q})] + \delta_{k'k}\gamma_{p'p}(\mathbf{q})\end{aligned} \tag{136}$$

where the designation of the angular integration variables has been left implicit as before, as have the diagonal requirement on J and M. The term $\hat{K}_H$ consists of the second group of terms from (125) (namely, the nuclear kinetic energy terms), and the last term in (136) is given by

$$\gamma_{p'p}(\mathbf{q}) = \frac{\hbar^2}{2}\Bigg\{\sum_{\alpha\beta}\langle\psi_{p'}|\hat{l}_\alpha\hat{l}_\beta|\psi_p\rangle_{\mathbf{z}}M_{\alpha\beta} + \sum_\alpha\langle\psi_{p'}|\hat{l}_\alpha|\psi_p\rangle_{\mathbf{z}}\lambda_\alpha$$
$$-\sum_{k,l=1}^{3H-6}G_{kl}\Bigg(\left\langle\psi_{p'}\left|\frac{\partial}{\partial q_k\,\partial q_l}\right|\psi_p\right\rangle_{\mathbf{z}}$$
$$+\left\langle\psi_{p'}\left|\frac{\partial}{\partial q_k}\right|\psi_p\right\rangle_{\mathbf{z}}\frac{\partial}{\partial q_l} + \left\langle\psi_{p'}\left|\frac{\partial}{\partial q_l}\right|\psi_p\right\rangle_{\mathbf{z}}\frac{\partial}{\partial q_k}\Bigg)$$
$$+\sum_{k=1}^{3H-6}\left[\frac{2}{i}\left\langle\psi_{p'}\left|(\mathbf{W}\hat{\mathbf{l}})_k\frac{\delta}{\delta g_k}\right|\psi_p\right\rangle_{\mathbf{z}} - \tau_k\left\langle\psi_{p'}\left|\frac{\partial}{\partial q_k}\right|\psi_p\right\rangle_{\mathbf{z}}\right]\Bigg\} \tag{137}$$

In (136) the role of $E_p(\mathbf{q}) + V(\mathbf{q})$ as a potential in the nuclear variables is clear.

The expression analogous to (70) can be developed in a similar fashion by using $\hat{K}_R(\boldsymbol{\phi}, \mathbf{q}, \mathbf{z})$, and the result is

$$\langle JMk'p'|\hat{K}_R|JMkp\rangle_{\mathbf{z}}$$
$$= \frac{\hbar^2}{4}(b_{+2}C^+_{Jk+1}C^+_{Jk}\,\delta_{k'k+2} + b_{-2}C^-_{Jk-1}C^-_{Jk}\,\delta_{k'k-2})\,\delta_{p'p}$$
$$+\frac{\hbar^2}{4}\{C^+_{Jk}[b_{+1}(2k+1) + \lambda_+]\,\delta_{k'k+1}$$
$$+ C^-_{Jk}[b_{-1}(2k-1) + \lambda_-]\,\delta_{k'k-1}\}\,\delta_{p'p}$$
$$+\frac{\hbar^2}{4}[C^+_{Jk}\gamma^+_{p'p}(\mathbf{q})\,\delta_{k'k+1} + C^-_{Jk}\gamma^-_{p'p}(\mathbf{q})\,\delta_{k'k-1}]$$
$$+\frac{\hbar^2}{2}\{[J(J+1) - k^2]b + b_0k^2 + \lambda_0 k\}\,\delta_{k'k}\,\delta_{p'p}$$
$$+\frac{\hbar^2}{2}\,\delta_{k'k}\,k\gamma^0_{p'p}(\mathbf{q}) \tag{138}$$

The meaning of the terms in this equation is exactly as in (70) but defined in relation to the nuclear variables alone. The γ terms simply extend the definition of the λ terms as in (130) and are specified by means of

$$\gamma^{\alpha}_{p'p}(\mathbf{q}) = 2\left(\langle \psi_{p'} | (\mathbf{M}\hat{\mathbf{l}})_{\alpha} | \psi_p \rangle_{\mathbf{z}} + \frac{1}{i} \sum_{k=1}^{3H-6} W_{k\alpha} \left\langle \psi_{p'} \left| \frac{\partial}{\partial q_k} \right| \psi_p \right\rangle_{\mathbf{z}} \right) \tag{139}$$

with γ^0 and $\gamma^{\pm}$ defined in terms of the γ^{α} in a manner analogous to the definition of the equivalent λ quantities in (72).

The dipole moment expression (73) may also be developed in the form (109) to give in place of (54)

$$\mathbf{d}(\mathbf{t}'', \mathbf{t}^e) = \mathscr{D}^{1*}\mathbf{d}(\mathbf{q}, \mathbf{z}) \tag{140}$$

with

$$\mathbf{d}(\mathbf{q}, \mathbf{z}) = e \sum_{i=1}^{H-1} \tilde{Z}_i \mathbf{X}^T \mathbf{z}_i^n(\mathbf{q}) - e\tilde{Z} \sum_{i=1}^{L} \mathbf{X}^T \mathbf{z}_i$$

or

$$\mathbf{d}(\mathbf{q}, \mathbf{z}) = \mathbf{d}(\mathbf{q}) - \mathbf{d}(\mathbf{z}) \tag{141}$$

and the matrix element on the right of (73) develops into

$$\langle \Phi'^{J'}_{k'p'} | \bar{d}_s(\mathbf{q}) | \Phi^J_{kp} \rangle \tag{142}$$

with

$$\bar{d}_s(\mathbf{q}) = d_s(\mathbf{q})\, \delta_{p'p} - \langle \psi_{p'} | d_s(\mathbf{z}) | \psi_p \rangle_{\mathbf{z}} \tag{143}$$

where the implied integral in (142) is over the nuclear variables $\mathbf{q}$, and the prime on the Φ on the left is used to indicate both that the nuclear motion functions can correspond to different internal energy states (vibrational states in the standard approach) and that each one of these states can be labeled by J, k, and p.

If this were as far as it was necessary to go in our account, a computational scheme would now be clear. A set of functions $\psi_p(\mathbf{q}, \mathbf{z})$ would be generated using the linear variation method on H^{elec} in a basis of trial functions chosen at a sequence of values of $\mathbf{q}$. In $\gamma_{p'p}$ and $\gamma^{\alpha}_{p'p}$ the coefficients that arise from integrating over the $\mathbf{z}$ variables would be calculated for each chosen value of $\mathbf{q}$, just as $E_p(\mathbf{q})$ is calculated, and these, together with the $E_p(\mathbf{q})$ values, would be extended into functions over all $\mathbf{q}$ perhaps by fitting to analytic forms. The effective nuclear motion problem specified by (136) and (138) could then be solved in the basis Φ^J_{nkp} (where n is used to label the components of the basis) to yield vibration–rotation energies and states. It would be hoped that, at least in the electronic ground state, it would be sufficient to use a single electronic state in (132) to account for most of the nuclear motion states of interest. If that were the case, then only a single potential would be needed and only diagonal elements of the γ matrices would arise. Then one could use a linear variation approach, for example, to the solution of the effective nuclear motion problem within any rotational manifold as specified by J.

However, equation (133) is not quite the problem that is usually solved in practical calculations. The practical solutions are specified by the clamped-nucleus Hamiltonian, and it remains to be discovered how that Hamiltonian is related to (133). It is also necessary, in view of what was said at the end of the previous section about the possible vanishing of the Jacobian, to investigate more closely the status of the product approximation in equation (132). These matters will be discussed in the next section, and in the course of the discussion an account will be given of the way in which permutational symmetry is to be incorporated.

4. The Clamped-Nucleus Hamiltonian and Permutational Symmetry

Section 3 showed how the body-fixed internal motion problem may be expressed in terms of two sets of coordinates and explained in what sense the $\mathbf{q}$ coordinates could be regarded as nuclear variables and the $\mathbf{z}$ coordinates as electronic variables. It showed further how the internal motion problem can be reduced to an effective problem specified in terms of the $\mathbf{q}$ coordinates alone if a manifold of functions $\psi_p(\mathbf{q}, \mathbf{z})$ with the properties of equations (134) and (135) can be specified.

It would be a standard tactic to assert that the manifold could be regarded as complete and hence that the exact solution could be expressed as an infinite sum of products of the form in equation (132). However, the remark made above about the possible vanishing of the Jacobian for particular values of the $\mathbf{q}$ indicates that such an assertion here would be more than usually problematic. At the very least it would be necessary to limit the assertion to a set of products that give rise to no divergent expectation values.

To complete our discussion, therefore, it is necessary to consider the status of the product functions and also to determine the position of the standard clamped-nucleus Hamiltonian in relation to the equations that have been derived in the previous section. In fact, it will prove convenient to present a combined account of these features that will lead naturally into a consideration of the role of permutational symmetry.

In practice, the potentials and electronic wave functions that are thought of as being used (and which are, in fact, used when nonempirical calculations are possible) arise not from solutions of (133) but as approximate solutions to the clamped-nucleus electronic problem specified as:

$$\hat{H}^{cn}(\mathbf{a}, \mathbf{x}^e) = -\frac{\hbar^2}{2m}\sum_{i=1}^{L} \nabla^2(\mathbf{x}_i^e) - \frac{e}{4\pi\varepsilon_0}\sum_{i=1}^{H}\sum_{j=1}^{L}\frac{Z_i}{|\mathbf{x}_j^e - \mathbf{a}_i|} + \frac{e^2}{8\pi\varepsilon_0}\sum_{ij=1}^{L}{}' \frac{1}{|\mathbf{x}_i^e - \mathbf{x}_j^e|} \tag{144}$$

This Hamiltonian is obtained from the laboratory-fixed one simply by assigning the values $\mathbf{a}_i$ to the nuclear variables $\mathbf{x}_i^n$; hence the designation *clamped-nucleus* for this form. The problem now is to decide in what way its solutions are related to those of (133).

The Hamiltonian (144) specifies a perfectly well-posed problem, and arguments such as are given in Simon[13] show that for any neutral or positively charged species the Hamiltonian has an infinite number of bound states (in the sense of being square integrable with negative eigenvalues) for arbitrary choice of a set of the $\mathbf{a}_i$. The eigenvalue for a particular choice of the $\mathbf{a}_i$ is called the electronic energy at the specified geometry.

Although the $\mathbf{a}_i$ in (144) are regarded simply as possible values of the $\mathbf{x}_i^n$ and hence chooseable at will, in practice in any sequence of calculations the sets are chosen so that no set in the sequence can be mapped on to any other set in the sequence merely by means of a uniform translation or a rigid rotation or rotation–inversion. Thus, the $\mathbf{a}_i$ are regarded as specifying definite nuclear *geometries* rather than absolute positions in space. The electronic energies calculated with these constraints on the $\mathbf{a}_i$ can therefore be regarded as representative values of a single-valued function $E(\mathbf{q})$ such as might be an eigenfunction in (133). (This to to avoid at this stage a difficulty that will be discussed later, namely, that it is usual to identify the nuclei in a clamped–nucleus calculation even if they are identical. This is clearly a questionable procedure in the context of a full calculation in which permutational symmetry is to be maintained.) Thus, with such constraints on sequences of the $\mathbf{a}_i$, $\hat{H}^{cn}$ can be thought of consistently as being of the same form as $\hat{H}^{\text{elec}}$. However, these considerations do not force particular choices for $\mathbf{C}$ or $\mathbf{q}$, so they do not force a particular form on $\hat{H}^{\text{elec}}$, and how the connection between this and the clamped-nucleus Hamiltonian is to be made is again a matter of judgment. From what has gone before, however, it seems very plausible that the correspondence should be made by thinking of (144) as related to that form of $\hat{H}^{\text{elec}}$ in which the center-of-nuclear-mass choice has been made together with the choice $a = 0$ in (93) and $\alpha = 1$ in (112). With these choices,

$$\hat{H}^{\text{elec}}(\mathbf{q}, \mathbf{z}) = -\frac{\hbar^2}{2\mu}\sum_{i=1}^{L}\nabla^2(\mathbf{z}_i) - \frac{e^2}{4\pi\varepsilon_0}\sum_{j=1}^{H-1}\sum_{j=1}^{L}\frac{Z_i}{f'_{ij}(\mathbf{q}, \mathbf{z})} + \frac{e^2}{8\pi\varepsilon_0}\sum_{ij=1}^{L}{}' \frac{1}{|\mathbf{z}_j - \mathbf{z}_i|} - \frac{\hbar^2}{2M}\sum_{ij=1}^{L}{}' \nabla(\mathbf{z}_i)\cdot\nabla(\mathbf{z}_j) \qquad (145)$$

with

$$\mu^{-1} = m^{-1} + M^{-1} \qquad (146)$$

and following (107)

$$f'_{ij}(\mathbf{q}, \mathbf{z}) = \left|\sum_{k=1}^{H-1} \mathbf{z}_k^n(\mathbf{q})(V^n)_{ki}^{-1} - \mathbf{z}_j\right| \qquad (147)$$

The implied restrictions on the $\mathbf{a}_i$ are such that, whatever the choice of $\mathbf{V}^n$ and the form of the $\mathbf{q}$ there is some set of values, denote it $\bar{\mathbf{q}}$, such that

$$f'_{ij}(\bar{\mathbf{q}}, \mathbf{z}) = |\mathbf{x}_j^e - \mathbf{a}_i| \qquad (148)$$

for any restricted choice of the $\mathbf{a}_i$. Thus, equation (144) can be identified with the first three terms in (145) by interpreting m as μ and $\mathbf{x}_i^e$ as $\mathbf{z}_i$. With this reinterpretation (144) can be regarded as approximating (145) simply by neglect of the last (so-called mass-polarization) term. Because M is always at least about 2000 times bigger than m it is not unreasonable to believe that, all other things being equal, the mass-polarization term is small and can be incorporated by perturbation theory on any set of approximate solutions of (144) at a later stage in the calculations.

Obviously, there are again no intrinsic reasons for thinking of trial functions for (144) in terms of any particular choice of $\mathbf{C}$ or any particular choice of $\mathbf{q}$. There is thus no particular form of (125) with which (144) ought to be associated—all forms are equally good—so that where the Jacobian vanishes is neither here nor there from the standpoint of $\psi_p(\mathbf{q}, \mathbf{z})$ because wherever it happens, this function remains well defined. Thus, to provide valid solutions for the full problem in terms of products of the form in equation (132) it must be the case, as asserted earlier, that it is the functions $\Phi_{kp}^J(\mathbf{q})$ that must be chosen to vanish strongly where the Jacobian vanishes.

For the reasons given at the end of Section 2, it is possible to choose, at least on a state-by-state basis, a body-fixed coordinate system in which a given space-fixed function can be approximated. Thus, it is possible to do so in the present form, because the part of the Jacobian that corresponds to the transformation to the $\mathbf{z}$ is just a constant, and it is only the $\mathbf{q}$ that contribute to the possible vanishing of the Jacobian. Experience of practical calculation is congruent with this view for it certainly seems possible in particular cases to produce by calculation accounts of rotation–vibration spectra over a wide compass of wavelengths that agree almost quantitatively with the results of experiment.

There remains, however, one further complication, which arises from the permutational symmetry of the problem, that must be dealt with before bringing the review to an end. It was noted in Section 2 that solutions of the full problem must correspond to irreducible representations of the permutation group of the relevant sets of identical particles. Not all irreducible representations are allowed, however, because the Pauli principle requires that wave functions for identical fermions (particles with odd half-integer spin) shall transform like the antisymmetric representation of the symmetric group and those for bosons (zero and integer spin) shall transform like the totally symmetric representation. For example, it can be shown (see Ref. 28, for example) that a collection of spin functions for spin one-half fermions (electrons or protons, for example) can provide a basis only for irreducible representations labeled by Young diagrams with at most two rows. Thus, to satisfy the Pauli principle the space functions must provide a basis for irreducible representations labeled by Young diagrams with at most two columns. Similar considerations apply to all the possible spin types. But it is not the precise form of the allowed function that matters; rather, it is the essential need to know precisely how a function actually does transform under the permutation of variables. In the case of a function expressed in body-fixed coordinates it turns out that in general the $\mathbf{q}$ coordinates are mixed only among themselves as are the $\mathbf{z}$ coordinates, but the $\boldsymbol{\phi}$ coordinates get mixed up with the $\mathbf{q}$. Thus, it is in general not possible to maintain the vibration–rotation separation implicit in (64) [or, equally, in equation (129)] because the function $|JMk\rangle$ will be changed from a function of the Euler angles only to a function of both Euler angles and internal coordinates.

What happens can perhaps be most clearly seen in terms of an example. Consider a triatomic composed of three identical particles, and consider just the nuclear variables. It is not necessary to consider explicitly the $\mathbf{z}$ coordinates because the way that they are defined ensures that they do not get mixed up with the $\boldsymbol{\phi}$ coordinates and that they transform in a completely standard fashion under permutations of the electrons and are invariant under permutations of the nuclei. Let the space-fixed coordinates be a pair of bond vectors

$$\mathbf{t}_1^n = \mathbf{x}_3^n - \mathbf{x}_2^n, \qquad \mathbf{t}_2^n = \mathbf{x}_1^n - \mathbf{x}_2^n \tag{149}$$

so that the inverse is

$$\begin{aligned} \mathbf{x}_1^n &= \mathbf{X} - \tfrac{1}{3}\mathbf{t}_1^n + \tfrac{2}{3}\mathbf{t}_2^n \\ \mathbf{x}_2^n &= \mathbf{X} - \tfrac{1}{3}\mathbf{t}_1^n - \tfrac{1}{3}\mathbf{t}_2^n \\ \mathbf{x}_3^n &= \mathbf{X} + \tfrac{2}{3}\mathbf{t}_1^n - \tfrac{1}{3}\mathbf{t}_2^n \end{aligned} \tag{150}$$

Let the internal coordinates be chosen as two bond lengths and the bond angle

$$r_1 = |\mathbf{t}_1^n|, \qquad r_2 = |\mathbf{t}_2^n|, \qquad \cos\theta = \frac{\mathbf{t}_1^{nT}\mathbf{t}_2^n}{r_1 r_2} \tag{151}$$

and suppose that the matrix $\mathbf{C}$ is chosen so that

$$\mathbf{z}_1^n = \mathbf{C}^T\mathbf{t}_1^n = \begin{pmatrix} 0 \\ 0 \\ r_1 \end{pmatrix}, \qquad \mathbf{z}_2^n = \mathbf{C}^T\mathbf{t}_2^n = r_2\begin{pmatrix} \sin\theta \\ 0 \\ \sin\theta \end{pmatrix} \tag{152}$$

If we think of the polar coordinates of $\mathbf{t}_1^n$ as ϕ_1 (range 0 to 2π) and ϕ_2 (range 0 to π), then

$$\begin{aligned} C_{xz} &= \frac{t_{x1}^n}{r_1} = \sin\phi_2\cos\phi_1 \\ C_{yz} &= \frac{t_{y1}^n}{r_1} = \sin\phi_2\sin\phi_1 \\ C_{zz} &= \frac{t_{z1}^n}{r_1} = \cos\phi_2 \end{aligned} \tag{153}$$

Thus, ϕ_1 and ϕ_2 are two of the Euler angles of the problem corresponding to rotations about the space-fixed z-axis and y-axis, respectively, in the standard choice of parameterization.[15]

If we now consider the permutations, which are represented in the space-fixed frame [see equation (97)], as

$$\mathbf{t}^{n'} = \mathbf{t}^{n}\mathbf{H} \tag{154}$$

then because the $\mathbf{t}^{n'}$ are merely a renaming of the original coordinates, in order to save writing primes we adopt the standard sleight-of-hand and denote the effect of the variable change in a function consequent upon a permutation by

$$\mathbf{t}_i^n \rightarrow \mathbf{t}^n\mathbf{H}^{-1} \tag{155}$$

Thus, in the present case the change in the internal coordinates r_i is

$$\begin{aligned} r_i &\rightarrow |\mathbf{t}_1^n H_{1i}^{-1} + \mathbf{t}_2^n H_{2i}^{-1}| \\ &= [r_1^2(H_{1i}^{-1})^2 + r_2^2(H_{2i}^{-1})^2 + 2H_{1i}^{-1}H_{2i}^{-1} r_1 r_2 \cos\theta]^{1/2} \end{aligned} \tag{156}$$

and similarly for θ.

To be specific, let us consider the transposition in equation (12), in which case

$$\mathbf{H} = \begin{pmatrix} 1 & 0 \\ -1 & -1 \end{pmatrix} = \mathbf{H}^{-1} \tag{157}$$

then

$$\begin{aligned} r_1 &\rightarrow (r_1^2 + r_2^2 - 2r_1 r_2 \cos\theta)^{1/2} \equiv \bar{r}_1, \qquad r_2 \rightarrow r_2 \\ \cos\theta &\rightarrow \frac{r_2 - r_1\cos\theta}{\bar{r}_1} \equiv \cos\bar{\theta} \end{aligned} \tag{158}$$

The bar notation is used simply as a shorthand and indicates that the changed variable is reexpressed in terms of the original variables. Similarly, one can write for the change induced by the permutation in the chosen column of $\mathbf{C}$:

$$C_{\alpha z} \rightarrow \frac{t_{\alpha 1}^n - t_{\alpha 2}^n}{\bar{r}_1} \equiv \bar{C}_{\alpha z} \tag{159}$$

It is clear that this column remains normalized under the induced change, and it is easy to show generally that $\bar{\mathbf{C}}$ is an orthogonal matrix. Because any orthogonal matrix can be written as a product of two orthogonal matrices one can always write

$$\bar{\mathbf{C}} = \mathbf{CU} \quad \text{with} \quad \mathbf{U} = \mathbf{C}^T\bar{\mathbf{C}} \tag{160}$$

and in the present example

$$U_{zz} = \sum_{\alpha} C_{\alpha z}\bar{C}_{\alpha z} = \sum_{\alpha} t^n_{\alpha 1}(t^n_{\alpha 1} - t^n_{\alpha 2})/\bar{r}_1 = \sum_{\alpha} z^n_{\alpha 1}(z^n_{\alpha 1} - z^n_{\alpha 2})/\bar{r}_1 \tag{161}$$

showing that U_{zz} is a function of the internal coordinates only. Had the other elements of $\mathbf{U}$ been examined, then they too would have been found to be of the same kind so that the change induced in $\mathbf{C}$ by the permutation can be written in the form

$$\mathbf{C}(\boldsymbol{\phi}) \rightarrow \mathbf{C}(\boldsymbol{\phi})\mathbf{U}(\mathbf{q}) \tag{162}$$

and this result is, in fact, a general one.

Now consider the change induced in the $\mathbf{z}^n$ by the permutation

$$\mathbf{z}^n = \mathbf{C}^T\mathbf{t}^n \rightarrow (\mathbf{CU})^T\mathbf{t}^n\mathbf{H}^{-1} \tag{163}$$

so that it follows at once that

$$\mathbf{z}^n \rightarrow \mathbf{U}^T\mathbf{z}^n\mathbf{H}^{-1} \tag{164}$$

is the equation to describe the changes induced in the $\mathbf{z}^n$, and clearly this result is general if (162) is general. A little thought shows, however, that this result actually guarantees the generality of the form given for the change induced in $\mathbf{C}$. It is certainly always possible to write the induced change in product form, but it is not immediately obvious that by choosing $\mathbf{C}$ to be one member of the product the other must have elements that are, at most, functions of the internal coordinates only. But had this not been the case, then clearly the change induced in the $\mathbf{z}^n$ would have produced coordinates that depended on the Euler angles. However, this is impossible by hypothesis because the construction of the $\mathbf{z}^n$ makes them invariant under orthogonal transformations, and thus their forms must be angle independent. It is possible, of course, that the matrix $\mathbf{U}$ can be a constant matrix, but in general it will depend on the internal coordinates.

It follows then, as stated above, that under a permutation changes are induced in the internal coordinates that can always be expressed in terms of the internal coordinates, whereas the changes induced in the angular coordinates produce functions that can depend on both the angular and the internal coordinates. Of course, because the expressions in the laboratory-fixed frame for all the operators that have been considered are obviously invariant under permutations of like particles, and because this invariance cannot be lost on passing to the body-fixed frame, the body fixed-operators show the same invariance. Indeed, it is possible, though extremely tedious, to show this explicitly. With coordinates chosen so that the requirements in equations (90)–(92) are satisfied, it is possible to show that the terms in brackets in the expression for the kinetic energy operator (125) are separately invariant, as are the three terms in the potential energy expression given by (128). Some care has to be taken in discussing the invariance of the potential in the uncoupled form of the Hamiltonian, and this will be dealt with in more detail later when considering how to use the clamped-nucleus electronic solutions, but otherwise the invariance will simply be asserted. However, the trial functions are not invariant, and it is the changes induced in them under permutations that allow them to be symmetry classified. It is this classification that must be made to ensure that any chosen function obeys the Pauli principle. Thus, we must consider the change induced by a permutation in

$$\Psi^{J,M}(\boldsymbol{\phi}, \mathbf{q}, \mathbf{z}) = \sum_{k=-J}^{+J} \Phi_k^J(\mathbf{q}, \mathbf{z})\,|JMk\rangle \tag{165}$$

It is not necessary to consider permutations of electrons because these are realized simply as permutations of the $\mathbf{z}_i$. However, the change induced by permutations of the nuclei on the $\mathbf{q}$ and $\boldsymbol{\phi}$ does require consideration in relation to (165). To begin this, consider first the form of $|1Mk\rangle$ as given by (48) and (49)

$$|1Mk\rangle = \left(\frac{3}{8\pi^2}\right)^{1/2} (\mathbf{X}^T\mathbf{C}\mathbf{X})_{Mk} \tag{166}$$

Because the elements of $\mathscr{D}^J$ can be obtained by repeated vector coupling of the elements of $\mathscr{D}^1$, a similar process is possible for the $|JMk\rangle$ from the $|1Mk\rangle$, and thus it is sufficient to know how $|1Mk\rangle$ transforms in order to know the general result. Using (162) for the change in **C** it follows from (166) that

$$
\begin{aligned}
|1Mk\rangle &\rightarrow \left(\frac{3}{8\pi^2}\right)^{1/2} (\mathbf{X}^T\mathbf{C}\mathbf{U}\mathbf{X})_{Mk} \\
&= \left(\frac{3}{8\pi^2}\right)^{1/2} (\mathbf{X}^T\mathbf{C}\mathbf{X}\mathbf{X}^\dagger\mathbf{U}\mathbf{X})_{Mk} \\
&= \sum_{n=-1}^{+1} |1Mn\rangle \mathscr{D}^1_{nk}(\mathbf{U})
\end{aligned}
\tag{167}
$$

so that the change induced in the general symmetric-top function under $\mathscr{D}$ is:

$$
|JMk\rangle \rightarrow \sum_{n=-J}^{+J} |JMn\rangle \mathscr{D}^J_{nk}(\mathbf{U}) \tag{168}
$$

In this equation $\mathscr{D}^J(\mathbf{U})$ is the matrix made up from the elements of **U** in exactly the same way that $\mathscr{D}^J$ is made up from the elements of **C**. A precise account of how this is to be done is given in Section 6.9 of Ref. 16. Should it turn out that **U** is a constant matrix, then $\mathscr{D}^J(\mathbf{U})$ is a constant matrix, and (168) simply represents a linear combination. If **U** is a unit matrix, then $|JMk\rangle$ is invariant. It should be noted here that this coupling of rotations by the permutations can mean that certain rotational states are not allowed by the Pauli principle, and this is important in assigning statistical weights to rotational states.

It is rather difficult to say anything precise about the change induced in the q_k under the permutation, because the change

$$
\mathbf{q}(\mathbf{z}^n) \rightarrow \mathbf{q}(\mathbf{U}^T\mathbf{z}^n\mathbf{H}^{-1}) \tag{169}
$$

has no general form and so the best that can be said is that a permutation of nuclei induces a general function change

$$
\Phi^J_k(\mathbf{q}, \mathbf{z}) \rightarrow \Phi'^J_k(\mathbf{q}, \mathbf{z}) \tag{170}
$$

where the precise nature of the function change depends on the permutation, the chosen form of the internal coordinates, and on the chosen functional form. Thus, the general change induced in (165) by $\mathscr{P}$ is

$$\Psi^{J,M}(\boldsymbol{\phi}, \mathbf{q}, \mathbf{z}) \rightarrow \sum_{k=-J}^{+J} \sum_{n=-J}^{+J} \mathscr{D}^{J}_{nk}(\mathbf{U})\Phi'^{J}_{k}(\mathbf{q}, \mathbf{z})\,|JMn\rangle$$

$$= \sum_{n=-J}^{+J} \bar{\Phi}^{J}_{n}(\mathbf{q}, \mathbf{z})\,|JMn\rangle \tag{171}$$

In practice, these results are of very little use because it is usually quite impossible to determine $\mathbf{U}(\mathbf{q})$ explicitly, and even if it were possible it would not be automatic, so that to do it for a symmetric group of any size would be a killing endeavor. However, such an endeavor is apparently necessary if the states allowed by the Pauli principle are to be determined, together with their statistical weights.

How to cope with permutational symmetry for a complete molecular wave function has been a vexing problem since the very beginning of molecular quantum mechanics. It has occasioned an enormous amount of work, particularly since the publication in 1963 of a paper by Longuet-Higgins[29] in which permutations were divided into *feasible* and *unfeasible* types, and in which it was argued that it was necessary to consider only the (often rather small) set of feasible permutations in a given problem. We shall return to a consideration of this division later. Implicit in almost all the discussions (including the one in Ref. 29 cited above) has been the assumption of the decoupling of electronic and nuclear motions. It has usually been thought sufficient to describe the nuclear motion as occurring within a single specified electronic state so that the nuclear motion Hamiltonian involves just a single potential. Usually implicit too (though sometimes not acknowledged) is the use of the Eckart Hamiltonian for nuclear motion, as this usage is the traditional one for spectroscopy in interpreting experimental results. It will be convenient for the purposes of the exposition here to follow the traditional path mapped out above. Much of the relevant work up to 1981 is reviewed in the monograph by Ezra,[17] and the following exposition relies heavily on it.

Remember that the Eckart Hamiltonian is constructed on the assumption that it is possible to define for a molecule an equilibrium geometry at which, it is assumed, the potential has a deep minimum. The body-fixed coordinates are then constructed in terms of this framework geometry so that if the instantaneous displacements of the particles from their equilibrium positions were zero, then perfect rotation–vibration separation would be achieved. It is achieved as closely as is theoretically possible for small displacements. Furthermore, the structure of the body-fixed Hamiltonian is such that the internal coordinates can be chosen as normal coordinates so that if the potential is well represented by a quadratic Taylor expansion about the minimum, then to a first approximation, each normal mode satisfies a standard simple harmonic oscillator equation. It is also the case that the inverse inertia tensor is, again to a first approximation, the inverse of the inertia tensor for the equilibrium geometry.

Thus, to a first approximation for small-amplitude internal motions and not too high J rotational motions, a suitable wave function is of the optimally separated form

$$\Psi^{JM}(\boldsymbol{\phi}, \mathbf{q}, \mathbf{z}) = \Phi(\mathbf{Q})\psi(\mathbf{0}, \mathbf{z}) \sum_{k=-J}^{+J} c_k^J |JMk\rangle \tag{172}$$

Here $\mathbf{Q}$ are the normal coordinates chosen so that $\mathbf{Q} = \mathbf{0}$ specifies the equilibrium geometry, and Φ is a product of simple harmonic oscillator functions, one for each normal coordinate. The function ψ is the electronic wave function used formally to specify the potential, evaluated at the equilibrium geometry. The c_k^J are coefficients determined by diagonalizing the angular momentum part of the Hamiltonian in the basis of symmetric-top functions. They are constants to first order because the inverse inertia tensor is a constant matrix to first order and because, also to first order, the Coriolis terms (often incorrectly called the vibrational angular-momentum terms in this context) vanish.

Because the nuclei are identified in order to specify the equilibrium geometry, the geometrical figure so specified can define a finite point group, and the structure of the Eckart Hamiltonian reflects that point group symmetry, if any. The normal coordinates are thus symmetry coordinates in the usual sense, with respect to the irreducible representations of the point group. The inverse inertia tensor and the Coriolis terms also reflect the point group symmetry so that it is reflected in the rotational part of the solutions.

Historically, it has been found sufficient in order to interpret the overwhelming majority of the experimental results of vibration–rotation spectroscopy to classify the states according to the point group symmetry of the equilibrium geometry by means of an Eckart-like scheme. The permutational symmetry of the problem is invoked only to get statistical weights and then only in a very special way that depends on relating the point group symmetry to that of the full symmetric group. (See, for example, Section 6.7 of Ref. 30.) This clearly is a rather odd state of affairs. It is as if there had been some sort of spontaneous symmetry breaking to lower the symmetry from that which might be anticipated from the standpoint of the permutational invariance to that of the point group invariance.

The transformation to the body-fixed frame in the Eckart approach is without approximation so that the full permutational symmetry of the kinetic energy operator is preserved. Clearly then, this apparent symmetry breaking must have its origin in the use of the actual, or presumed, clamped-nucleus potential. As explained above, the potential in equation (128) is invariant term-by-term under permutations of like nuclei and of electrons. In particular, $V(\mathbf{q}, \mathbf{z})$ as used in (133) is invariant under the permutation of like nuclei. Because the $\mathbf{q}$ in equation (133) are simply parameters it follows that this invariance must be carried through to $E_p(\mathbf{q})$. Thus, if the potential used in the Eckart Hamiltonian actually corresponded to one derived from an equation like (133), one would expect the full permutational invariance to be preserved. That this is not generally the case arises from the fact that full permutational invariance is not generally preserved in solutions of the clamped-nucleus problem, be they actual or presumed, because the nuclei are always identified, and nothing extra is done by appropriate symmetrization to remove this identification.

The puzzle is why this clearly defective approach actually seems to provide a satisfactory account of much molecular spectra. There is, at present, no clarification of this puzzle that is found satisfactory by all, so it would be inappropriate to speculate further here. It is possible, however, to make matters a little clearer, at least within the Eckart approach. Louck and Galbraith[(31)] noticed that in considering permutational symmetry within the Eckart approach there were certain permutations of identical particles which if considered to operate on the framework geometry had the same effect as one of the point group operations similarly regarded. Such an operation they called a perrotation. Ezra[(17)] refined and developed this notion and showed that for perrotations the matrix $\mathbf{U}(\mathbf{q})$ had constant elements so that (162) simply specifies a linear combination and the vibration–rotation separation of (172) is thus effectively preserved. For the ammonia molecule *all* the permutations of like particles are perrotations, but this is not the case in ethene. If in ethene an allowed permutation which is not a perrotation is made, then the change in (172) according to (162) is simply to make the internal coordinate part of the function J and k dependent. Because such terms do not appear to matter, however, it is assumed that they are of very small amplitude in the region of interest. However, in the absence of detailed examination (which would be pretty tricky), it is not possible to say anything certainly.

The last part of the above discussion has been on the basis of the Eckart approach, in which only point group invariance of the potential has been assumed. It is clear from the general discussion that the potential in which any decoupled nuclear motion problem ought really to be solved should be invariant under the full permutation group of the identical nuclei. This means that in order to generate such a potential, it is necessary to do, or at least imagine done, clamped-nucleus calculations that allow for every permutation of the identified but identical nuclei. But this clearly is not only a daunting task (there are 518,400 operations to be considered in benzene, for instance), but it is also inconsistent with the idea of an equilibrium geometry.

No one does this sort of thing in practice, and an attempt at explaining why not is given in Ref. 29 by saying that it is not possible to perform certain permutations because to do so would involve crossing an insuperable energy barrier. All other permutations are called feasible permutations, and these clearly include the perrotations. What permutations are feasible, it is argued, is a matter for theoretical and experimental investigation on a case-by-case basis.

Useful though this approach may be in limiting the extent of our anguish, it cannot really be a fundamental account of what is involved. The idea of permuting identical particles is a mathematical one and not a physical one, and there can be no idea of energy barriers and the like until the parts of the Hamiltonian are uncoupled and then only in the simplest product approximation for the uncoupled wave function. Certainly it would be impossible to advance the feasibility arguments in terms of the fully coupled problem, where the permutation requirements are just the same. In the future it might prove possible to provide an account in terms of the solutions to the full problem, without uncoupling or introducing potential surfaces, which would explain the apparent symmetry breaking as far as the point-group level as arising from some sort of clustering of states or a kind of accidental degeneracy. But so far such an explanation eludes us. Thus in calculations and in interpreting spectra it is important to proceed cautiously and on a case-by-case basis to make sure that we do not make egregious errors. The idea of feasible permutations seems to provide as reasonable a guide to what can be done as anything that is at present available. An extended account of what is involved here together with some rules for feasible operations can be found in the book by Bunker.[32] Some attempts to press at the limits of this approach can be found collected, often in summary form, in the volume edited by Maruani and Serre,[33] and some of the problems that arise in attempting global descriptions in body-fixed coordinate systems are discussed in a review article by Natanson.[34]

5. Conclusion

It has been the aim of this chapter to try to offer, at least at a formal level, a rational scheme by which the electronic and nuclear motion parts of quantum-chemical calculations can be brought together. The autonomous role of the clamped-nucleus electronic Hamiltonians has been exhibited, and it has been shown that its solutions can plausibly be mapped onto more than one form of the body-fixed electronic Hamiltonian but that there are good reasons for preferring the center-of-nuclear-mass operator given in (145) to any other form. The essentially local nature of any body-fixed Hamiltonian for nuclear motion has been exhibited as an inevitable consequence of coordinate regions in which the Jacobian matrix of the transformation to the body-fixed system becomes singular. Finally, an account has been given of the problems that lie in the way of providing a proper account of the permutational symmetry associated with identical particles in constructing trial wave functions.

Appendix 1. Alternative Forms for the Jacobian and the Hamiltonian

The matrix of the Jacobian invoked in (65) as arising from (34) and (35) may be written as

$$\mathbf{J} = \bar{\mathbf{C}}(\mathbf{\Omega}\,|\,\mathbf{Q})\bar{\mathbf{D}} \tag{A1.1}$$

In this equation $\bar{\mathbf{C}}$ is a block diagonal matrix with $N-1$ repeats of the matrix $\mathbf{C}$ along the diagonal, and $\bar{\mathbf{D}}$ is similarly block diagonal with $\mathbf{D}$ forming the first block and the unit matrix E_{3n-6} forming the second. In the partitioned matrix the elements of $\mathbf{\Omega}$ are the $\mathbf{\Omega}^i$ treated as 3×3 blocks to form a column, and the elements of $\mathbf{Q}$ are similarly formed from the $\mathbf{Q}^i$ treated as $3\times 3N-6$ blocks. The matrix of the inverse Jacobian arising from (38) and (39) may be written in partitioned form as

$$\mathbf{J}^{-1} = \bar{\mathbf{D}}^{-1}(\hat{\mathbf{z}}\,|\,\check{\mathbf{Q}})^T\mathbf{C}^T \tag{A1.2}$$

The elements of the partitioned matrix here are formed just as above but from the $\hat{\mathbf{z}}^i$ and the $\check{\mathbf{Q}}^i$, respectively. The internal coordinate part of the metric arising from (A1.1) can thus be written as

$$\begin{pmatrix} \mathbf{M} & \mathbf{W}^T \\ \mathbf{W} & \mathbf{G} \end{pmatrix} \tag{A1.3}$$

where the elements of the partitioned matrix are defined in (58), (59), and (61).

The internal coordinate part of the inverse metric arising from (A1.2) can likewise be written as

$$\begin{pmatrix} \mathbf{I} & \mathbf{Y} \\ \mathbf{Y}^T & \mathbf{F} \end{pmatrix} \tag{A1.4}$$

To define the component matrices of the inverse metric it is necessary to define the matrix $\boldsymbol{\mu}$ inverse to the matrix defined in (12). It is seen easily that its elements are

$$\mu_{ij} = \sum_{k=1}^{N} m_k(\mathbf{V}^{-1})_{ik}(\mathbf{V}^{-1})_{jk}, \qquad i, j = 1, 2, \ldots, N-1 \tag{A1.5}$$

The component matrices are

$$\mathbf{I} = \sum_{i,j=1}^{N-1} \mu_{ij}\hat{\mathbf{z}}_i^T\hat{\mathbf{z}}_j \tag{A1.6}$$

and this matrix is clearly the body-fixed form of the instantaneous inertia tensor, whereas

$$\mathbf{Y} = \sum_{i,j=1}^{N-1} \mu_{ij}\hat{\mathbf{z}}_i^T\tilde{\mathbf{Q}}^j, \qquad \mathbf{F} = \sum_{i,j=1}^{N-1} \mu_{ij}\tilde{\mathbf{Q}}^{iT}\tilde{\mathbf{Q}}^j \tag{A1.7}$$

Because the metric matrices above are inverse to each other, standard matrix manipulations enable it to be shown that

$$\begin{aligned} \mathbf{F}^{-1} &= (\mathbf{G} - \mathbf{W}\mathbf{M}^{-1}\mathbf{W}^T), \qquad & \mathbf{I}^{-1} &= (\mathbf{M} - \mathbf{W}^T\mathbf{G}^{-1}\mathbf{W}) \\ \mathbf{G}^{-1} &= (\mathbf{F} - \mathbf{Y}^T\mathbf{I}^{-1}\mathbf{Y}), \qquad & \mathbf{M}^{-1} &= (\mathbf{I} - \mathbf{Y}\mathbf{F}^{-1}\mathbf{Y}^T) \end{aligned} \tag{A1.8}$$

and that

$$\mathbf{W}\mathbf{M}^{-1} + \mathbf{F}^{-1}\mathbf{Y}^T = \mathbf{0}_{3N-6,3} \tag{A1.9}$$

Using these results it is possible to rewrite the kinetic energy operator (57) in a form analogous to that found by Eckart[6] by introducing the Coriolis coupling operator

$$\hat{\boldsymbol{\pi}} = \frac{\hbar}{i}\mathbf{M}^{-1}\mathbf{W}^T\frac{\partial}{\partial \mathbf{q}} \tag{A1.10}$$

so that (57) becomes

$$\hat{K}(\boldsymbol{\phi}, \mathbf{q}) = \frac{1}{2}\left[\sum_{\alpha\beta} M_{\alpha\beta}(\hat{L}_\alpha + \hat{\pi}_\alpha)(\hat{L}_\beta + \hat{\pi}_\beta) + \hbar \sum_\alpha \nu_\alpha \hat{L}_\alpha\right]$$

$$-\frac{\hbar^2}{2}\left(\sum_{k,l=1}^{3N-6} F_{kl}^{-1} \frac{\partial^2}{\partial q_k\, \partial q_l} + \sum_{k=1}^{3N-6} \tau'_k \frac{\partial}{\partial q_k}\right) \qquad \text{(A1.11)}$$

Here

$$\tau'_k = \tau_k + \sum_{l=1}^{3N-6} W_{k\beta} \frac{\partial}{\partial q_k} (\mathbf{M}^{-1}\mathbf{W}^T)_{\beta l} \qquad \text{(A1.12)}$$

and all the other components not defined above are exactly as defined in connection with (57). It can be shown[35] that if the Eckart embedding is chosen for the body-fixed system, then (A1.12) above reduces to Louck's form of the Eckart–Watson kinetic energy operator.[18]

There is no need, if it is not convenient, to pass to the internal coordinates through an explicit choice of space-fixed coordinates, but if the direct path is chosen, then it is necessary explicitly to consider the center-of-mass coordinate in constructing the Jacobian. This means that the matrix $\bar{\mathbf{C}}$ now has an extra diagonal block of $\mathbf{C}$, whereas the unit matrix in $\bar{\mathbf{D}}$ is now increased in dimension to $\mathbf{E}_{3N-3}$. The internal coordinate part of the Jacobian now has an extra $3N \times 3$ partition which consists of the $N3 \times 3$ matrices, $M_T^{-1} m_i \mathbf{E}_3$ for $i = 1, 2, \ldots, N$, and there is a similar extension to the inverse Jacobian consisting of N repetitions of $\mathbf{E}_3$. The requirement that these matrices be inverse to each other means that the internal coordinates must be chosen so that the following relationships are satisfied as identities:

$$\sum_{i=1}^{N} \boldsymbol{\Omega}^i = \mathbf{0}_3, \qquad \sum_{i=1}^{N} \mathbf{Q}^i = \mathbf{0}_{3,3N-6}$$

$$\sum_{i=1}^{N} m_i \mathbf{z}_i = \mathbf{0}_3, \qquad \sum_{i=1}^{N} m_i \tilde{\mathbf{Q}}^i = \mathbf{0}_{3,3N-6} \qquad \text{(A1.13)}$$

These requirements extend somewhat those of the set (40), and the equivalent extension of (41) is

$$\boldsymbol{\Omega}^i \hat{\mathbf{z}}_j + \mathbf{Q}^i \tilde{\mathbf{Q}}^j = \delta_{ij} \mathbf{E}_3 - M_T^{-1} m_i \qquad \text{(A1.14)}$$

The metric and its inverse are now constructed using $m_i^{-1}\,\delta_{ij}$ in place of μ_{ij}^{-1} and $m_i\,\delta_{ij}$ in place of μ_{ij}, respectively. The metric matrix then becomes $3N \times 3N$ with an extra 3×3 block on the diagonal with elements M_T^{-1}. Similarly, the inverse metric has an extra 3×3 block with elements M_T. Only null blocks connect these to the rest of the matrix, and the components of the rest of the matrix generalize in an obvious way with sums over particle indices running to N rather than $N-1$. The form of the kinetic energy operator in equation (57) is preserved as is the Eckart-like form given above.

Appendix 2. The Role of Parity Using $O(3)$ as the Invariance Group

It was remarked at the end of Section 2 that in the body of this review the rotation–inversion symmetry of the problem has been factored according to $SO(3)$ rather than the full group $O(3)$. This appendix is concerned with the appropriate extension.

The group $O(3)$ is the direct product of the inversion group C_i with the special orthogonal group in three dimensions, $SO(3)$. The inversion group consists of the identity operator $\hat{E}$ and the inversion operator $\hat{I}$. The operations of $SO(3)$ may be realized in three-dimensional coordinate space by proper 3×3 orthogonal matrices $\mathbf{R}$, and the inversion may be realized by $-\mathbf{E}_3$. Thus, for every matrix $\mathbf{R}$ in $SO(3)$ there is a companion matrix $-\mathbf{R}$ in $O(3)$. The matrix $-\mathbf{R}$ in general represents a reflection such that if a proper rotation by π is performed about the normal to the reflection plane, and this operation is followed by the inversion, then the matrix $-\mathbf{R}$ results. It has been noted already [see equation (49) et seq.] that it is possible to construct irreducible representations, $\mathscr{D}^j$, by vector coupling of the elements of $\mathscr{D}^1$, a matrix that can be derived directly from the orthogonal matrices themselves. From this it is easy to see that, given a proper rotation $\mathbf{R}$,

$$\mathscr{D}^J(-\mathbf{R}) = (-1)^J \mathscr{D}^J(\mathbf{R}) \tag{A2.1}$$

where the usage for $\mathscr{D}^J$ is explained in the discussion following equation (168).

Thus, the matrices constructed in this way provide irreducible representations of even parity for J even but of odd parity for J odd. In a colloquial sense at least, they provide only half the representations there should be for $O(3)$ because there should also be representations of odd parity for J even and of even parity for J odd because the group manifold consists of two disconnected but isomorphic sheets. This defect can be remedied, however, by following the work of Biedenharn and Louck[16] and of Ezra[17] and defining

$$\mathscr{D}^{0J}(\mathbf{R}) = \mathscr{D}^{J}(\mathbf{R}), \qquad \mathscr{D}^{1J}(\mathbf{R}) = |\mathbf{R}|\mathscr{D}^{J}(\mathbf{R}) \tag{A2.2}$$

where $|\mathbf{R}|$ denotes the determinant of $\mathbf{R}$. It follows that

$$\mathscr{D}^{rJ}(-\mathbf{R}) = (-1)^{r+J}\mathscr{D}^{rJ}(\mathbf{R}), \qquad r = 0, 1 \tag{A2.3}$$

If the angular momentum eigenfunctions given by (48) are generalized to

$$|JMkr\rangle = |\mathbf{C}|^{r}|JMk\rangle \tag{A2.4}$$

then it is easily shown that under the change $\mathbf{C} \to \mathbf{CU}$ equation (168) generalizes to

$$|JMkr\rangle = \sum_{n=-J}^{+J} |JMnr\rangle\mathscr{D}^{rJ}_{nk}(\mathbf{U}) \tag{A2.5}$$

This establishes the position of the matrices $\mathscr{D}^{rJ}$ as representation matrices for the general orthogonal transformation in three dimensions. In particular, if $\mathbf{U} = -\mathbf{E}_3$, then $\mathscr{D}^{rJ}(-\mathbf{E}_3) = (-1)^{r+J}\mathbf{E}_3$ as is required. It further establishes that in the definition of $\mathscr{D}^{1}$ the sign of the determinant of the orthogonal matrix is immaterial to the validity of the definition.

The underlying integral for the normalization of $|JMkr\rangle$ must now extend over both sheets of $O(3)$ and is thus[16]

$$\int \mathscr{D}^{r'J'*}_{M'k'}(\mathbf{C})\mathscr{D}^{rJ}_{Mk}(\mathbf{C})|\mathbf{D}|^{-1}\,d\phi + \int \mathscr{D}^{r'J'*}_{M'k'}(-\mathbf{C})\mathscr{D}^{rJ}_{Mk}(-\mathbf{C})|\mathbf{D}|^{-1}\,d\phi \tag{A2.6}$$

where $|\mathbf{D}|^{-1}$ is the angular part of the Jacobian as explained in the discussion following equation (67), and the integrals are each over the whole range of the Euler angles. If $r' = r$, then both integrals are the same and have the value $\delta_{J'J}\,\delta_{M'M}\,\delta_{k'k}8\pi^2/(2J+1)$, whereas if $r' \neq r$ then both integrals have the same absolute value but opposite signs and hence they cancel. Thus, the generalized rotation–inversion eigenfunctions are orthonormal according to

$$\langle J'M'k'r' | JMkr \rangle = \delta_{J'J}\,\delta_{M'M}\,\delta_{k'k}\,\delta_{r'r} \tag{A2.7}$$

In a discussion of inversion and parity toward the end of Chapter 19 of his book *Group Theory*, Wigner[36] notes that in two- and three-body systems (and in these systems only) it is possible to realize the effect of an inversion by means of a sequence of proper rotations because each of these cases have rather special features. Thus, the matrix **C** that is used to transform to the body-fixed system can be chosen in the three-body case such that all three particles lie in a plane. If that plane is, say, the $x - y$ one, then the z coordinates of the particles will all be zero and the effect of an inversion can be achieved by performing a proper rotation by π about the z-axis. Two-body systems have not been considered in this review for reasons explained earlier, but three-body systems have not so far been excluded. However, the only genuinely three-body systems of interest are such systems as the helium atom and the hydrogen molecule ion, and these are best dealt with by quite special methods that are specific to each problem. Thus, three-body systems can also be ignored without much loss of generality. To do so is not to ignore triatomic molecules for in these systems there will be four or more particles, because at least one electron must be present for binding. Thus, the foregoing discussion can be regarded as the general treatment of parity for present purposes.

In the discussion of internal coordinates following equation (26) it was noted that internal coordinates composed from scalar triple products of space-fixed coordinates would not be used in the present work because it was desired that the internal coordinates should be invariant not only under rotations but under rotation–inversions too. Not everyone makes that restriction; for example, it is not made in the work of Shipsey,[37] but if it is not made then a distinctly different discussion parity from that above must be given.

Considering the effect of inversion on the operators discussed in Section 2, it is easy to see that inversion leaves the kinetic energy, the potential energy, and the angular momentum operators invariant. The dipole moment operator, however, has an induced sign change. In the body-fixed form of this operator this change is carried by the sign change in $\mathscr{D}^1$ as $\mathbf{C} \to -\mathbf{C}$. Thus, equation (54) should be rewritten with $\mathscr{D}^1$ replaced by $\mathscr{D}^{01}$.

It follows then that (69) generalizes to

$$\langle J'M'k'r' | \hat{K}_I + V | JMkr \rangle = \delta_{r'r}\,\delta_{J'J}\,\delta_{M'M}\,\delta_{k'k}(\hat{K}_I + V) \tag{A2.8}$$

and that (70) generalizes in a similar manner given that the $|JMkr\rangle$ remain eigenfunctions of the angular momentum operators just as in (45), irrespective of the r value.

Equation (56) generalizes as

$$\langle J'M'k'r'|\mathscr{D}^{01*}_{mp}|JMkr\rangle$$
$$= (1 - \delta_{ss'})(-1)^{M'+k}[(2J'+1)(2J+1)]^{1/2}$$
$$\times \begin{pmatrix} J' & 1 & J \\ -M' & m & M \end{pmatrix} \begin{pmatrix} J' & 1 & J \\ k' & p & -k \end{pmatrix} \quad \text{(A2.9)}$$

where the parity s is $(-1)^{J+r}$ and similarly for r', and thus (73) generalizes accordingly. When the initial and final states are the same state, then the formula yields the expectation value of the dipole for that state. However, because the parity factors in the wave functions are the same and the parity of the dipole operator is odd, the expectation value must be zero, so it would seem that no molecule can have a permanent dipole moment—a somewhat counterintuitive result.

Before trying to elucidate this further, consider the expected value of the Hamiltonian operator. From what has been shown above it is clear that any pair of states that differ just in r value will constitute an energy degenerate pair. This degeneracy must be "accidental", however, in the sense that it is not due to $O(3)$ symmetry. This is because $O(3)$ is the direct product $C_i \times SO(3)$ so that there will be two distinct irreducible representations of $O(3)$ for every representation of $SO(3)$. This is precisely analogous to the way in which distinct g and u representations arise in point groups such as C_{6h} or D_{6h} which are distinct products $C_i \times C_6$ or $C_i \times D_6$. The fact that the two representations are distinct means that there is no group theoretical reason to suppose that the two states $|JMkr\rangle$ for $r = 0, 1$ *should* be degenerate.

It is perfectly possible that a state with a particular r value will be disallowed by the Pauli principle. Whether or not that is the case would have to be determined in any particular occurrence by the changes induced according to (171), suitably generalized by replacing $\mathscr{D}^J$ with $\mathscr{D}^{rJ}$. As explained in the discussion of that equation, this would be exceptionally tricky. However, Ezra[17] has discussed the problem in detail in some special cases. But if states with both r values were allowed, then it would be possible to understand the presence of a permanent dipole moment by attributing it to the mixing of the degenerate states by the small perturbing electric field used for the measurement. The observed dipole moment would then be analogous to that observed in the excited states of the hydrogen atom, which is customarily attributed to the mixing by the electric field of the accidentally degenerate *s* and *p* orbitals.

References

1. F. Heitler and F. London, *Z. Physik* **44**, 455–472 (1927).
2. S. C. Wang, *Phys. Rev.* **31**, 579–586 (1928).
3. M. Born and J. R. Oppenheimer, *Ann. der Phys.* **84**, 457–484 (1927).
4. J. M. Combes and R. Seiler, in *Quantum Dynamics of Molecules* (R. G. Woolley, ed.), pp. 435–482, Plenum, New York (1980).
5. J. M. Combes, *Acta Physica Austriaca, Supp. XVII*, 139–159 (1977).
6. C. Eckart, *Phys. Rev.* **47**, 552–558 (1935).
7. J. K. G. Watson, *Mol. Phys.* **15**, 479–490 (1968).
8. R. J. Whitehead and N. C. Handy, *J. Mol. Spec.* **55**, 356–373 (1975).
9. G. D. Carney and R. N. Porter, *J. Chem. Phys.* **60**, 4251–4264 (1974).
10. M. Born and K. Huang, Appendix 8 in *Dynamical Theory of Crystal Lattices*, Oxford University Press, Oxford, England (1954).
11. J. C. Slater, *Proc. Nat. Acad. Sci.* **13**, 423–430 (1927).
12. W. Kolos and L. Wolniewicz, *J. Chem. Phys.* **41**, 3663–3678 (1964).
    D. M. Bishop and L. M. Cheung, *Phys. Rev. A* **18**, 1846–1852 (1978).
13. B. Simon, *Quantum Mechanics for Hamiltonians Defined as Quadratic Forms*, Princeton University Press, Princeton, NJ (1971).
14. B. T. Sutcliffe and J. Tennyson, *Int. J. Quant. Chem.* **39**, 183–196 (1991).
15. D. M. Brink and G. R. Satchler, *Angular Momentum*, 2nd ed., Clarendon Press, Oxford (1968).
16. L. C. Biedenharn and J. C. Louck, *Angular Momentum in Quantum Physics*, Addison-Wesley, Reading, MA. (1982).
17. G. Ezra, *Symmetry Properties of Molecules*, Lecture Notes in Chemistry **28**, Springer-Verlag, Berlin (1982).
18. J. C. Louck, *J. Mol. Spec.* **61**, 107–137 (1976).
19. E. C. Kemble, *The Fundamental Principles of Quantum Mechanics*, McGraw-Hill, New York (1937).
20. B. T. Sutcliffe, in *Theoretical Models of Chemical Bonding* (Z. Maksic, ed.), Pt. 1, pp. 1–28, Springer-Verlag, Berlin (1990).
21. R. Bartholomae, D. Martin, and B. T. Sutcliffe, *J. Mol. Spec.* **87**, 367–381 (1981).
22. J. Tennyson and B. T. Sutcliffe, *J. Mol. Spec.* **101**, 71–82 (1983).
23. S. Carter and N. C. Handy, *Mol. Phys.* **53**, 1033–1039 (1984).
24. N. C. Handy, *Mol. Phys.* **61**, 207–233 (1987).
25. F. T. Smith, *Phys. Rev. Lett.* **45**, 1157–1160 (1980).
26. G. Hunter, *Int. J. Quant. Chem.* **9**, 237–242 (1975).
27. J. Czub and L. Wolniewicz, *Mol. Phys.* **36**, 1301–1308 (1978).
28. M. Hamermesh, *Group Theory and Its Application to Physical Problems*, Addison-Wesley, Reading, MA (1962).
29. H. C. Longuet-Higgins, *Molec. Phys.* **6**, 445–460 (1963).
30. I. G. Kaplan, *Symmetry of Many-Electron Systems*, Academic, London (1975).
31. J. D. Louck and H. W. Galbraith, *Rev. Mod. Phys.* **48**, 69–106 (1976).
32. P. R. Bunker, *Molecular Symmetry and Spectroscopy*, Academic, London (1975).

33. J. Maruani and J. Serre, *Symmetries and Properties of Nonrigid Molecules*, Elsevier, Amsterdam (1983).
34. G. A. Natanson, *Advan. Chem. Phys.* **58**, 55–126 (1985).
35. B. T. Sutcliffe, in *Quantum Dynamics of Molecules* (R. G. Woolley, ed.), pp. 1–37, Plenum, New York (1980).
36. E. P. Wigner, *Group Theory*, Academic, New York (1959).
37. E. J. Shipsey, *J. Chem. Phys.* **91**, 4813–4820 (1989).

3

The Calculation of Highly Excited Vibration–Rotation States of Triatomic Molecules

Jonathan Tennyson, Steven Miller, and James R. Henderson

1. Introduction

For quantum chemists, the 1970s saw tremendous advances in the solution of the molecular electronic structure problem. For the first time it became possible to do ab initio calculations of chemical accuracy on systems of chemical interest. This development was jointly driven by advances in computer technology and improved techniques of treating, in particular, electron correlation effects.

Although improvements in electronic structure calculations are naturally continuing, in many ways the 1980s was the decade of nuclear dynamics. To a large extent, the impetus behind this was not improved computer technology but the availability of reliable potential energy surfaces, both ab initio and semiempirical. These have provided a strong stimulus for improving methods of treating nuclear motions. In particular, variational techniques have proved very powerful for obtaining bound states of small,

Jonathan Tennyson, Steven Miller, and James R. Henderson • Department of Physics and Astronomy, University College London, London WC1E 6BT, U.K.

Methods in Computational Chemistry, Volume 4: Molecular Vibrations, edited by Stephen Wilson. Plenum Press, New York, 1992.

particularly triatomic, systems. The situation with regard to early variational calculations was ably reviewed by Carney et al.[1] for chemically bound systems and Le Roy and Carley[2] for Van der Waals complexes.

Work of particular note covered by the reviews included the Whitehead–Handy method,[3,4] which is still the variational method of choice for solving the normal coordinate, the Eckart–Watson Hamiltonian[5,6]; the variational, internal coordinate calculations performed by Lai and Hagstrom[7]; and the use of numerically defined basis functions by Le Roy and Van Kranendonk.[8] We will return to all these methods later in this chapter.

Since 1980, there has been a burgeoning of methods for treating the vibrational motions of small molecules. A number of these methods have been reviewed in a series of articles in *Computer Physics Reports*.[9–11] In the early 1980s a number of calculations on floppy systems, such as KCN[12–14] and CH_2^+,[15–21] showed that there were practical problems in using the Eckart–Watson Hamiltonian for such systems. The difficulty was not only the well-known singularity for linear geometries but also that in the Whitehead–Handy method, and no doubt in other algorithms too, it was easy to leave the true domain of the problem (see, for example, Ref. 22).

This led to the development of a series of formally exact, body-fixed Hamiltonians defined in internal molecular coordinates; examples are given in Refs. 7, 14, 18, and 23–25. Sutcliffe[26] derived an algorithm for obtaining these Hamiltonians, discussed in Chapter 2 of this book, which Handy has since automated using the algebraic language REDUCE.[27] The disadvantage with this approach is that different coordinates, and thus different Hamiltonians, are required for the optimal treatment of each class of vibrational problem. Additionally, these internal coordinate Hamiltonians also contain singularities which need careful treatment.

Our approach, for example, has relied heavily on the use of atom–diatom scattering, or Jacobi, coordinates for triatomic systems.[10,28,29] This approach has proved robust because it is always possible to avoid singularities.[14,17] But we would be the first to admit that these coordinates are not optimal for all three-atom systems.[17,30,31]

A solution to this problem, at least for triatomic systems, appears to have been found with the development of Hamiltonians in terms of flexible coordinate systems.[32–37] These coordinates enter the Hamiltonian only via the moment of inertia terms. This means that they can be left as user-defined parameters in a general computer code.[38,39] In the most extreme case the exact definition of the coordinates can even be used as a variational parameter in the calculation.[40] We believe that the most general set of triatomic internal coordinates are the ones recently proposed by Sutcliffe and Tennyson.[37] These will be discussed in detail in Section 2.

Given a Hamiltonian, a wide variety of solution strategies has been tried to obtain vibrational energy levels and wave functions. Except for simple systems (i.e., ones for which one can find coordinates in which the problem is nearly separable), the most accurate calculations on small systems have been performed variationally using basis set expansions. There is, of course, considerable art in how one chooses appropriate basis functions for a given calculation. This problem is discussed in Section 3.

Recently, Light and co-workers[25,41–46] have demonstrated how a particular finite-element method, the discrete variable representation (DVR), yields highly excited vibrational states with great efficiency. The DVR uses the insight given by conventional basis sets and at the same time gives a hierarchy of diagonalizations and truncations which lead to a final secular problem of greatly reduced dimension. Our recent implementation of the DVR approach[31,47,48] will be discussed in Section 3.

For systems with significant Coriolis coupling, it is not possible to separate completely the vibrational and rotational motions. The resulting coupling of vibrational and rotational basis functions, generally via the angular functions of the vibrational coordinate, leads to secular problems which grow rapidly with total angular momentum, J. With this, algorithm calculations with $J = 4$[30] were the limit unless other approximations were made.[49]

Independently, Chen et al.[22] and Tennyson and Sutcliffe[50] proposed two-step variational methods to circumvent the rotational problem. The essence of their methods, which differ in detail, is that basis functions for the fully coupled ro-vibrational problem are expanded in terms of solutions of some suitable "vibrational" problem. These "vibrational" basis functions are so efficient that calculations with J greater than 60 have now been performed.[46,51] As will be discussed in Section 4, the rotational motion problem can thus be regarded as effectively solved.

Of course, spectroscopy yields more than just energy levels or, more precisely, transition frequencies. The intensity of a particular transition yields information on both the molecular wave functions involved and the property surface, usually the dipole, that couples the states. This intensity data can act as a probe of the physical conditions of the molecules, a factor that is of particular importance for astrophysics. Although intensity calculations have been performed for a number of years,[52,53] it is only comparatively recently that several workers[54–57] have developed general programs for the computation of ro-vibrational dipole transition intensities. This work, our part of which is outlined in Section 5, proved timely in that it played an important role in the first-ever assignment of an extraterrestrial spectrum of H_3^+.[58]

The calculation of ro-vibrational wave functions and spectra has been greatly helped by the advances in computer technology. The secular equation or basis function methods are easily expressed in terms of matrix operations and are thus particularly suited to vector-processing machines. Section 6 is devoted to some of the algorithms that we have used to obtain efficient programs.

It is fair to regard the variational method for calculating the ro-vibrational spectra of triatomic systems from a given potential energy surface as a mature technique. Black box computer suites are available for performing these calculations,[11,39] and many systems have been studied. In Section 7 we review applications to certain typical classes of systems. The chapter finishes with concluding remarks in Section 8.

2. The Hamiltonian

2.1. Coordinate Systems

The Hamiltonian for a system with N bodies has $3N$ coordinates. However, the relative positions of these N bodies only require $3N-6$ so-called internal coordinates for a full description.

Starting from the initial $3N$ laboratory coordinates, one first has to remove the three coordinates associated with the translational motion of the center-of-mass. This is necessary to separate the discrete ro-vibrational spectrum from the translational continuum. The resulting $3N-3$ coordinates are usually called space-fixed.

It is usual to partition the bound-state motions of molecules in terms of rotations and vibrations. If this is to be done, one has to define three (two for linear molecules) coordinates which can be associated only with the rotational motion of the system, leaving $3N-6$ internal coordinates. Such coordinate sets are called body-fixed.

Body-fixed coordinates have a number of advantages for ro-vibrational calculations. They give a natural distinction between vibrational and rotational motion in line with the usual interpretation of molecular spectra. The integrals, particularly over angular coordinates, are considerably simpler in body-fixed than in space-fixed coordinates. Perhaps most importantly, body-fixed coordinates lend themselves naturally to the two-step variational procedure for calculating rotational excitation, discussed in Section 4.1.

Of course, it is possible to suggest a large number of plausible internal coordinate systems even for small-molecule systems. Deriving the body-fixed Hamiltonian for a given set of internal coordinates and orientation of the internal Cartesian axis system is not easy. Furthermore, there is no guarantee that the resulting Hamiltonian is actually going to be usable. This is a consequence of the coordinate transformations which inevitably introduce singularities into the body-fixed Hamiltonian.

Sutcliffe[26] has derived a general procedure for obtaining body-fixed Hamiltonians. This procedure has been used to obtain triatomic Hamiltonians for specific coordinate systems.[14,18,23] It has since been generalized to give whole classes of Hamiltonians by using parameterized coordinate sets.[32,37] Details of this procedure can be found in Chapter 2 of the present volume.

The most recent and most general triatomic Hamiltonian derived by Sutcliffe and Tennyson[32,37] is expressed in terms of any two geometrically defined distances and an included angle. Thus, if the position vector of atom i is $\mathbf{x}_i$, then these coordinates can be specified by writing

$$\begin{aligned}\mathbf{t}_1 &= a_1(\mathbf{x}_1 - \mathbf{x}_3) + b_1(\mathbf{x}_2 - \mathbf{x}_1) + c_1(\mathbf{x}_3 - \mathbf{x}_2)\\ \mathbf{t}_2 &= a_2(\mathbf{x}_1 - \mathbf{x}_3) + b_2(\mathbf{x}_2 - \mathbf{x}_1) + c_2(\mathbf{x}_3 - \mathbf{x}_2)\end{aligned} \tag{1}$$

for any choice of a_1, b_1, c_1, a_2, b_2, and c_2 for which $\mathbf{t}_1$ and $\mathbf{t}_2$ are linearly independent. The resulting internal coordinates are

$$r_1 = |\mathbf{t}_1|, \qquad r_2 = |\mathbf{t}_2| \quad \text{and} \quad \theta = \cos^{-1}\left(\frac{\mathbf{t}_1 \cdot \mathbf{t}_2}{r_1 r_2}\right) \tag{2}$$

These coordinates are so general that only a subset of them have actually been used.[37,39] This subset is specified by setting

$$\begin{aligned}a_1 = -g_2, \quad b_1 = 0, \quad c_1 = -1; \qquad 0 \le g_2 \le 1\\ a_2 = 1, \quad b_2 = 0, \quad c_2 = g_1; \qquad 0 \le g_1 \le 1\end{aligned} \tag{3}$$

Figure 1 illustrates these coordinates.

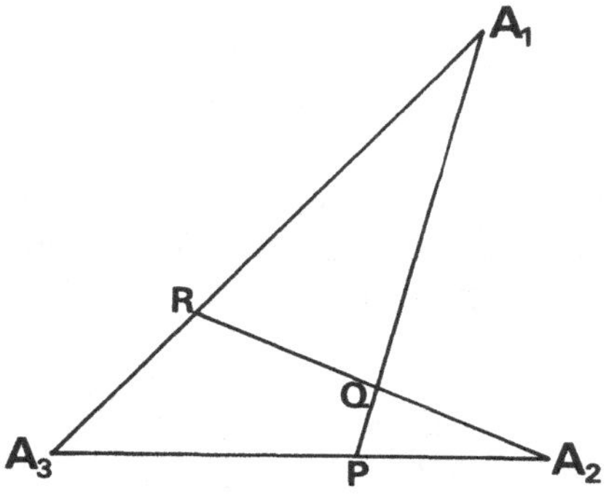

Figure 1. Coordinate system. A_i represents atom i. The coordinates in the text are given by $r_1 = A_2 - R$, $r_2 = A_1 - P$ and $\theta = A_1\hat{Q}A_2$. The geometric parameters are defined by $g_1 = (A_3 - P)/(A_3 - A_2)$ and $g_2 = (A_3 - R)/(A_3 - A_1)$.

Table 1. Special Cases of the General Sutcliffe–Tennyson Coordinate System[a]

Coordinate type[b]	g_1	g_2	Ref.
Scattering or Jacobi	$\dfrac{m_2}{m_2+m_3}$	0	10, 14
Bond length–bond angle	0	0	7, 11, 18
Geometric midpoint	1/2	0	32
Radau[c]	$1-\dfrac{\alpha}{\alpha+\beta-\alpha\beta}$	$1-\dfrac{\alpha}{1-\beta+\alpha\beta}$	44, 59

[a] From Ref. 37. Parameters g_1 and g_2 are defined in Fig. 1. Atom i has mass m_i. The references are to examples of the use of these coordinates for (ro-)vibrational calculations.
[b] In this definition neither the Jacobian nor Radau coordinates are mass weighted.
[c] $\alpha=[m_3/(m_1+m_2+m_3)]^{1/2}$, $\beta=m_2/(m_1+m_2)$.

With suitable choices for the parameters g_1 and g_2, one can obtain a number of well-known coordinate systems. These are summarized in Table 1.

In terms of these coordinates the body-fixed kinetic energy operator may be written as[37]

$$\hat{K}=\hat{K}_V^{(1)}+\hat{K}_V^{(2)}+\hat{K}_{VR} \tag{4}$$

where*

$$\hat{K}_V^{(1)}=-\frac{\hbar^2}{2}\left[\frac{1}{\mu_1}\frac{\partial^2}{\partial r_1^2}+\frac{1}{\mu_2}\frac{\partial^2}{\partial r_2^2}+\left(\frac{1}{\mu_1 r_1^2}+\frac{1}{\mu_2 r_2^2}\right)\left(\frac{1}{\sin\theta}\frac{\partial}{\partial\theta}\sin\theta\frac{\partial}{\partial\theta}\right)\right] \tag{5}$$

$$\hat{K}_V^{(2)}=\frac{\hbar^2}{\mu_{12}}\left[-\cos\theta\frac{\partial^2}{\partial r_1\,\partial r_2}+\left(\frac{1}{r_1}\frac{\partial}{\partial r_2}+\frac{1}{r_2}\frac{\partial}{\partial r_1}-\frac{1}{r_1r_2}\right)\left(\sin\theta\frac{\partial}{\partial\theta}+\cos\theta\right)\right.$$
$$\left.+\frac{\cos\theta}{r_1r_2}\left(\frac{1}{\sin\theta}\frac{\partial}{\partial\theta}\sin\theta\frac{\partial}{\partial\theta}\right)\right] \tag{6}$$

and $\hat{K}_{VR}$ is the operator that carries the rotational motion as well as the Coriolis coupling terms. This operator, which is null for the vibration only ($J=0$) problem, will be discussed in the following subsection.

* This form of the Hamiltonian is often called the manifestly Hermitian form. That is, the factor $r_1^2r_2^2$ has been eliminated from the Jacobian. This process is equivalent to choosing an internal coordinate wave function $r_1^{-1}r_2^{-1}\,{}^J\Phi_{k,n}$ and obtaining an operator for ${}^J\Phi_{k,n}$ alone.[37]

In the above equations the reduced masses are given by

$$\mu_i^{-1} = (a_i - b_i)^2 m_1^{-1} + (b_i - c_i)^2 m_2^{-1} + (c_i - a_i)^2 m_3^{-1}, \qquad i = 1, 2$$
$$\mu_{12}^{-1} = (a_1 - b_1)(a_2 - b_2)m_1^{-1} + (b_1 - c_1)(b_2 - c_2)m_2^{-1} \tag{7}$$
$$+ (c_1 - a_1)(c_2 - a_2)m_3^{-1}$$

or, for the coordinates defined in Fig. 1,

$$\mu_1^{-1} = g_2^2 m_1^{-1} + m_2^{-1} + (1 - g_2)^2 m_3^{-1}$$
$$\mu_{12}^{-1} = (1 - g_1)(1 - g_2)m_3^{-1} - g_2 m_1^{-1} - g_1 m_2^{-1} \tag{8}$$
$$\mu_2^{-1} = m_1^{-1} + g_1^2 m_2^{-1} + (1 - g_1)^2 m_3^{-1}$$

There is a particular class of coordinates that eliminate all cross-derivative terms from the kinetic energy operator. These coordinates are called orthogonal coordinates. Within the formalism given here they consist of the class of coordinates that satisfy the condition

$$\mu_{12}^{-1} = 0 \tag{9}$$

Obviously, this condition is sufficient to make $\hat{K}_V^{(2)}$ a null operator. Coordinate systems that are orthogonal include the scattering and Radau coordinates of Table 1.

Orthogonal coordinates are particularly important, probably essential, for finite-element methods such as the DVR. This is because in these methods the matrix elements of the kinetic energy operator are the major off-diagonal coupling elements in the Hamiltonian.[48] Conversely, in more conventional basis set methods the potential is the main source of off-diagonal coupling terms and nonorthogonal coordinate systems, such as bond length–bond angle coordinates, are often employed.[7,11]

2.2. Axis Embedding

Choosing a set of internal coordinates does not uniquely define a body-fixed Hamiltonian. For cases with nonzero total angular momentum, J, it is also necessary to specify a set of Cartesian axes fixed to the frame of the molecule. For the special case of $J = 0$ the Hamiltonian is invariant to this embedding, and in our coordinates the kinetic energy operator consists only of the vibrational terms given above.

The Euler angles linking the body-fixed and space-fixed axis systems are the three coordinates associated with the rotational motion of the whole molecule. We will call these angles (α, β, γ). The full Hamiltonian is a function of these angles. However, J is a constant of motion for the systems and the rotation matrices $D^J_{kM}(a, \beta, \gamma)$[60] form a complete set for a given J. It is thus usual to let the Hamiltonian act on D^J_{kM}: multiply from the left by $D^{J*}_{k'M}$ and integrate over α, β, γ. This yields an effective Hamiltonian which is only a function of the internal coordinates of the system.

In the above analysis, the quantum number M is the projection of J onto the space-fixed z-axis. Because the Hamiltonian is independent of M in the absence of a magnetic field, it will not be considered further. The quantum number k is the projection of the total angular momentum J onto the body-fixed z-axis and is in general not a constant of motion. However, neither the vibrational operators, $\hat{K}^{(1)}_V$ and $\hat{K}^{(2)}_V$, nor the potential couple terms of different k.

Sutcliffe and Tennyson,[37] in developing their generalized Hamiltonian, derived an effective vibration–rotation kinetic energy operator with the body-fixed z-axis embedded at an arbitrary angle of $a\theta$ from $\mathbf{t}_1$. The other axes were defined such that the molecule was in the x–z plane and the axes formed a right-handed set. In their formalism, a could take any value between 0 and 1. However, only for the two special cases of $a = 0$ and $a = 1$ were they able to remove completely the singularities in the operator. We will thus confine our discussion to these cases only.*

For the special case of embedding the z-axis parallel to one of the radial coordinates, a judicious choice of angular functions was found to remove all the angular singularities from the Hamiltonian.[14,37] This effective Hamiltonian is thus appropriate for both linear and bent molecules and, more importantly for us, is also suitable for treating floppy systems for which both linear and strongly bent configurations are important. The exact form of the Hamiltonian will be discussed in the following section.

3. Vibrational Excitation

3.1. Angular Basis Functions

In 1960, Arthurs and Dalgarno[61] used a partial wave expansion in the angle θ and the analytic properties of the angular coordinates to turn a scattering coordinate Hamiltonian into an effective Hamiltonian operating on only the radial coordinates. Following this general strategy one can define

* The singularity in the special case $a = 1/2$ is confined to $\theta = \pi$. This form of the Hamiltonian has also been used.[187]

angular basis functions

$$|j, k\rangle = \Theta_{j,k}(\theta) D^J_{kM}(\alpha, \beta, \gamma) \tag{10}$$

where $\Theta_{j,k}(\theta)$ is a normalized associated Legendre polynomial with the Condon and Shortley[62] phase conventions.

Letting the Hamiltonian act on $|j, k\rangle$, multiplying from the right by $\langle j', k'|$, and integrating over the angular variables *only* yields an effective radial Hamiltonian[32,37]

$$\hat{H}(r_1, r_2) = \hat{K}_V^{(1)} + \hat{K}_V^{(2)} + \hat{K}_{VR}^{(1)} + \hat{K}_{VR}^{(2)} + \delta_{k'k}\langle j', k| V(r_1, r_2, \theta)|j, k\rangle_\theta \tag{11}$$

In this expression V is the potential function and

$$\hat{K}_V^{(1)} = \delta_{j'j}\,\delta_{k'k}\left[-\frac{\hbar^2}{2\mu_1}\frac{\partial^2}{\partial r_1^2} - \frac{\hbar^2}{2\mu_2}\frac{\partial^2}{\partial r_2^2} + \frac{\hbar^2}{2} j(j+1)\left(\frac{1}{\mu_1 r_1^2} + \frac{1}{\mu_2 r_1^2}\right)\right] \tag{12}$$

$$\hat{K}_V^{(2)} = -\delta_{j'j+1}\,\delta_{k'k} d_{jk}\frac{\hbar^2}{2\mu_{12}}\left(\frac{\partial}{\partial r_1} - \frac{j+1}{r_1}\right)\left(\frac{\partial}{\partial r_2} - \frac{j+1}{r_2}\right)$$
$$- \delta_{j'j-1}\,\delta_{k'k} d_{j-1,k}\frac{\hbar^2}{2\mu_{12}}\left(\frac{\partial}{\partial r_1} + \frac{j}{r_1}\right)\left(\frac{\partial}{\partial r_2} + \frac{j}{r_2}\right) \tag{13}$$

$$\hat{K}_{VR}^{(1)} = -\delta_{j'j}\,\delta_{k'k}\frac{\hbar^2}{2\mu_1 r_1^2}[J(J+1) - 2k^2] - \delta_{j'j}\,\delta_{k'k\pm1}\frac{\hbar^2}{2\mu_1 r_1^2} C^\pm_{Jk} C^\pm_{jk} \tag{14}$$

$$\hat{K}_{VR}^{(2)} = \delta_{j'j+1}\,\delta_{k'k\pm1}\frac{\hbar^2}{2\mu_{12}} C^\pm_{Jk}\frac{a_{j\pm k}}{r_1}\left(\frac{j+1}{r_2} - \frac{\partial}{\partial r_2}\right)$$
$$+ \delta_{j'j-1}\,\delta_{k'k\pm1}\frac{\hbar^2}{2\mu_{12}} C^\pm_{Jk}\frac{b_{j\pm k}}{r_1}\left(\frac{j}{r_2} + \frac{\partial}{\partial r_1}\right) \tag{15}$$

The angular factors in the above equations are

$$C^{\pm}_{Jk} = [J(J+1) - k(k \pm 1)]^{1/2} \tag{16}$$

$$d_{jk} = \left[\frac{(j-k+1)(j+k+1)}{(2j+1)(2j+3)}\right]^{1/2} \tag{17}$$

$$a_{jk} = \left[\frac{(j+k+1)(j+k+2)}{(2j+1)(2j+3)}\right]^{1/2} \tag{18}$$

$$b_{jk} = \left[\frac{(j-k)(j-k-1)}{4j^2-1}\right]^{1/2} \tag{19}$$

The $\hat{K}_{VR}$ operators given above are for the z-axis embedded parallel to r_1 (i.e., $a = 0$). The appropriate operators for the $a = 1$ embedding are obtained by making the exchanges $r_1 \leftrightarrow r_2$ and $\mu_1 \leftrightarrow \mu_2$.

3.2. The Potential

The above discussion focused on a systematic treatment of the kinetic energy operators for particular coordinate systems. It is important to remember, however, that it is the potential energy (hyper)surface that is actually the key term in deciding the behavior of a particular system.[63]

The effective Hamiltonian of equation (11) involves integrating potential matrix elements over the angular coordinates. If the potential is expressed as an expansion in terms of Legendre polynomials

$$V(r_1, r_2, \theta) = \sum_{\lambda} V_\lambda(r_1, r_2)\Theta_{\lambda,0}(\theta) \tag{20}$$

then the angular integration can also be performed analytically

$$\langle j', k'|\Theta_{\lambda,0}(\theta)|j, k\rangle = \delta_{k'k}(-1)^k[(2j'+1)(2j+1)]^{1/2} \times \begin{pmatrix} j' & \lambda & j \\ 0 & 0 & 0 \end{pmatrix}\begin{pmatrix} j' & \lambda & j \\ -k & 0 & k \end{pmatrix} \tag{21}$$

In equation (21), the expressions in large brackets are $3-j$ symbols,[60] which make up what is known as a Gaunt coefficient.

Of course, most potentials are not conveniently expressed as Legendre expansions. However, it is possible to adapt an arbitrary analytically defined potential function to the above scheme using Gauss–Legendre quadrature.[64] This is done[17,29] by forming the expansion numerically for each (r_1, r_2) required by the radial integration schemes (see Section 3.3 below). This method can be shown to be a computationally very efficient way of forming matrix elements over the potential.[65]

3.3. Radial Basis Functions

The traditional approach to solving the effective Hamiltonian given in Section 3.1 is by direct solution of coupled integro-differential equations. In bound-state nuclear motion problems, the direct method has found particular favor for Van der Waals systems.[70] This is because the very different nature of the internal and intramolecular stretching coordinates allows different solution strategies to be employed for these modes. Thus, the ro-vibrational motions of an atom–diatom complex may be represented in Jacobian or scattered coordinates, and the vibrational motion of the diatomic, r_1, if not frozen, may be carried by a small basis set expansion. Conversely, very accurate solutions for the intramolecular stretching coordinates, r_2, can be obtained by direct integration. For further discussion see the recent review by Hutson.[66]

Rather than using direct numerical solution of the effective Schrödinger equations, we have preferred to use basis set expansions for both radial coordinates. At least in part this choice is the only feasible one for systems where the radial coordinates have to be treated on a near equal footing. A number of different radial basis sets have been employed.

Le Roy and co-workers[2,8] used numerically defined radial basis functions for their work on Van der Waals dimers. For certain simple systems, in particular, complexes involving molecular hydrogen, these functions are stunningly efficient.[67] However, in applications to more strongly bound and more strongly coupled systems, such as KCN[13] or LiCN,[49] it is more difficult to find a suitable model potential with which to define these numerical functions. Furthermore, even with current supercomputer technology, working with numerically defined functions in more than one dimension is difficult. We note however that Cropek and Carney[68] and Searles and von Nagy-Felsobuki[69] have used numerical functions in all three dimensions to solve for relatively rigid triatomic systems.

The Morse potential[70] is known to give a realistic and moderately flexible representation of a typical molecular stretching coordinate. Solutions of this potential, Morse oscillators, might therefore seem a natural choice for radial basis functions. However, there are some technical problems that need to be solved if one wants to work with these functions. Because the Morse potential only has a finite number of bound states, the Morse oscillators do not form a complete set without the inclusion of continuum functions. Furthermore, the Morse oscillators are defined in terms of associated Laguerre polynomials with indices coupled in a fashion that makes the functions difficult to work with. Finally, the Morse oscillators do not obey the correct boundary conditions at the limit $r \to 0$,[71] which can cause problems in some applications.[17]

Tennyson and Sutcliffe[14,72] circumvented the completeness problem by defining a set of Morse oscillatorlike functions

$$|n\rangle = H_n(r) = \beta^{1/2} N_{n\alpha} \exp\left(-\frac{x}{2}\right) x^{(\alpha+1)/2} L_n^{\alpha}(x)$$
$$x = A \exp[-\beta(r - r_e)] \tag{22}$$

where

$$A = \frac{4D_e}{\beta}, \qquad \beta = \omega_e \left(\frac{\mu}{2D_e}\right)^{1/2} \tag{23}$$

In the above, $N_{n\alpha} L_n^{\alpha}$ is a normalized associated Laguerre polynomial,[73] and μ is the reduced mass of the relevant coordinate. This corresponds to solutions of the Morse potential

$$V(r) = D_e\{1 - \exp[-\beta(r - r_e)^2]\} \tag{24}$$

if $\alpha = A - (2n + 1)$. Instead, Tennyson and Sutcliffe decided to work with functions defined by fixing A equal to the integer value of α. These functions, the lowest of which corresponds closely to the ground state of the Morse potential, form a complete orthonormal set and belong to a single set of polynomials.

In principle, the parameters that define the Morse potential, r_e, ω_e, and D_e, can be associated respectively with the equilibrium separation, fundamental frequency, and dissociation energy of the relevant coordinate. In practice, (r_e, ω_e, D_e) are treated as variational parameters and optimized accordingly. These optimizable functions have proved very successful for a whole variety of problems. Unless otherwise stated, all the applications discussed below used Morse oscillatorlike functions for the radial coordinates.

As already mentioned, Morse oscillators (and the Morse oscillatorlike functions) do not behave satisfactorily at $r = 0$. This is not a problem for coordinates that represent explicitly the distance between two nuclear centers because the vibrational wave function is vanishingly small at this limit. However, for other coordinates, such as r_2 in scattering coordinates, this may be physically accessible for some systems. In atom–diatom scattering coordinates the $r_2 = 0$ geometry corresponds to a linear geometry with the atom inserted at the center-of-mass of the diatom. For systems where such behavior is significant, alternative basis functions have to be found.

Tennyson and Sutcliffe[17,29] suggested the use of spherical oscillators for situations where the $r = 0$ limit is important. These may be defined by

$$|n\rangle = H_n(r) = 2^{1/2}\beta^{3/4}N_{n\alpha}\exp\left(-\frac{x}{2}\right)x^{\alpha/2}L_n^{\alpha+(1/2)}(x)$$
$$x = \beta r^2 \tag{25}$$

where

$$\beta = (\mu\omega_e)^{1/2} \tag{26}$$

and (α, ω_e) are treated as variational parameters.

Spherical oscillators have been used successfully for a number of calculations,[17,31] but they have generally been found to be less efficient than the Morse oscillatorlike functions when these are also viable.[30] Spherical oscillators have an additional disadvantage. For (quasi-) linear systems the value of α depends on the state being considered.[74] Thus, when the system has amplitude at $r = 0$, for instance in its rotational ground state, then α must be chosen equal to zero. But when the same system is then rotationally excited, removing the amplitude from the region, the optimum value of α increases. This is not only inconvenient, it leads to severe problems in calculating rotational constants for heavy systems by performing calculations for several rotational states. This is because the usual cancellation of convergence errors which occurs when each calculation is performed with the same basis functions no longer occurs.

For a so-called[42] finite basis representation (FBR) we are now in a position to write down the unsymmetrized expression for the wave function of the system. In terms of the basis functions described above, the approximation to the sth energy level, E_s^J has a wave function

$$\Psi_s^J = \sum_k \sum_{jmn} d_{kjmn}^{Js}|j, k\rangle|m\rangle|n\rangle \tag{27}$$

where $|m\rangle$ and $|n\rangle$ are the radial basis functions associated with the r_1 and r_2 coordinates, respectively.

In this approach the variational coefficients, d_{kjmn}^{Js}, are determined by diagonalizing the appropriate secular matrix. How exactly the matrix elements of this matrix are constructed will be dealt with in Section 4, but we first will consider an alternative way of representing the vibrational wave function of the problem.

3.4. Discrete Variable Representation

Use of a discrete variable representation (DVR) is not particularly new. It was originally suggested by Harris et al.[75] in the 1960s, when it was shown by Dickinson and Certain[76] to have formal, and useful, equivalences to Gaussian quadrature approximations. However, work using the DVR method, and other finite-element techniques,[77] has taken a sudden upsurge following a series of studies by Light and co-workers. This work has recently been reviewed by Bačić and Light.[45]

It appears that the most significant of a number of developments made using DVR schemes is that they provide a rigorous and physically motivated hierarchy of problems that can be diagonalized and the basis set truncated before solving the next problem. This results in relatively small final secular matrices which have many eigenvalues converged with respect to adding further intermediate functions. Typically about 40% of the DVR eigenvalues are converged as opposed to less than 10% in most FBR calculations. As such, the DVR technique bears a similarity to the approach of Carter and Handy[78] who prediagonalize (or contract) their raw basis sets using suitable model problems, often cutting through the potential. The difference is that in the DVR this contraction is done in an optimal manner on the actual problem.

The particular strength of the DVR approach is in obtaining large numbers of converged vibrational levels. Where comparisons have been made the DVR appears capable of yielding almost an order of magnitude more converged levels than the FBR with similar computational resources.

Not unnaturally, therefore, much of our recent work has been concerned with adapting our methodology to a DVR framework. Although DVRs have been used successfully in two or three dimensions,[25,79] most of the work has only used this methodology for a single coordinate. Here we will therefore simply consider the use of a one-dimensional DVR and refer to the review of Bačić and Light[45] for a more general discussion.

As mentioned above, the DVR is really only useful for orthogonal coordinates (ones for which $\mu_{12}^{-1} = 0$). This is because in the finite-element representation generated by the DVR transformation, all the off-diagonal terms are assumed to come from the kinetic energy operators. Inclusion of the extra off-diagonal terms that arise in nonorthogonal coordinates—for example, see $\hat{K}_V^{(2)}$ above—would make this method computationally inefficient.[48] The DVR will therefore only be developed for orthogonal coordinates (i.e., $\hat{K}_V^{(2)}$ will be assumed to be zero). In practice, this means scattering and Radau coordinates within our coordinate set (see Table 1), although other orthogonal coordinates are possible.[26]

If one starts from the effective Hamiltonian of equation (11), one can apply a DVR in the θ coordinate by defining a transformation based on N-point Gauss-associated Legendre quadrature. The quadrature points and weights are $\chi_k(=\cos \boldsymbol{\theta}_k)$ and $\boldsymbol{\omega}_k$. It should be noted that a different quadrature and hence a different transformation is generated for each value of $|k|$ and that there is no need to use the same number of DVR points for each k.

In terms of these quadrature points the transformation can be written

$$T_{ja}^{k} = \omega_{ka}^{1/2} \Theta_{jk}(\theta_{ka}) \tag{28}$$

where Θ_{jk} is assumed to be normalized. This transformation applied to the effective Hamiltonian (11) gives

$$\begin{aligned}
\hat{H}_{k',a',ka} &= \sum_{j=k}^{N+k-1} \sum_{j'=k'}^{N'+k'-1} T_{j'a'}^{k'} \hat{H}(r_1, r_2) T_{ja}^{k} \\
&= \delta_{k'k}\,\delta_{a'a} \left\{ \frac{-\hbar^2}{2\mu_1} \frac{\partial^2}{\partial r_1^2} - \frac{\hbar^2}{2\mu_2} \frac{\partial^2}{\partial r_2^2} \right. \\
&\quad \left. + \frac{\hbar^2}{2\mu_1 r_1^2} [J(J+1) - 2k^2] + V(r_1, r_2, \theta_{ka}) \right\} \\
&\quad + \delta_{k'k} \frac{\hbar^2}{2} \left(\frac{1}{\mu_1 r_1^2} + \frac{1}{\mu_2 r_2^2} \right) L_{a'a}^{k} - \delta_{k'k\pm 1} \frac{\hbar^2}{2\mu_1 r_1^2} C_{Jk}^{\pm} Q_{a'a}^{k\pm}
\end{aligned} \tag{29}$$

for the $a = 0$ embedding, where

$$L_{a'a}^{k} = \sum_{j=k}^{N+k-1} T_{ja'}^{k} j(j+1) T_{ja}^{k} \tag{30}$$

and

$$Q_{a'a}^{k\pm} = \sum_{j} T_{ja'}^{k\pm 1} C_{jk}^{\pm} T_{ja}^{k} \tag{31}$$

In the above sum, j runs from the maximum of $(k, k\pm 1)$ to the minimum of $(N+k-1, N'+k\pm 1-1)$.

In deriving $\hat{H}_{k'\alpha',k\alpha}$, the standard DVR quadrature approximation has been assumed:

$$\sum_{j,j'=k}^{N+k-1} T^{k}_{j'\alpha'} \langle j'k|V(r_1, r_2, \theta)|jk\rangle_\theta T^{k}_{j\alpha} \simeq \delta_{\alpha\alpha'} V(r_1, r_2, \theta_{k\alpha}) \tag{32}$$

This approximation is equivalent to using N-point Gauss-associated Legendre quadrature.[76] Solving this problem is equivalent to using an FBR including all Θ_{jk} up to $j = N + k - 1$. Because of the quadrature approximation, DVR methods are not strictly variational. However, for all but the smallest problems, solutions are in practice found to converge monotonically from above.

The solution strategy for the DVR method is to then solve the two-dimensional Hamiltonian obtained for each α

$$\hat{h}^{(2D)}_{k\alpha}(r_1, r_2) = \frac{-\hbar^2}{2\mu_1}\frac{\partial^2}{\partial r_1^2} - \frac{\hbar^2}{2\mu_2}\frac{\partial^2}{\partial r_2^2} + \frac{\hbar^2}{2\mu_1 r_1^2}[J(J+1) - 2k^2] + V(r_1, r_2, \theta_{k\alpha}) \tag{33}$$

The solutions of $\hat{\mathbf{h}}^{(2D)}$ are then used as a basis to solve the full three-dimensional problem. The advantage of this technique is that not all solutions of the two-dimensional problem are needed to converge the lower-lying three-dimensional wave functions. This leads to large savings in the size of the final secular matrix that needs to be diagonalized.

3.5. Symmetry

It is not strictly necessary to use the full symmetry of a system when performing variational calculations. However, doing so not only makes the calculations computationally more efficient but also eases the task of making suitable assignments to calculated states. Indeed, for one particular system (D_2S),[80] we found that neglecting symmetry led to an unphysical redistribution of the transition intensity between states which were of different symmetry but effectively degenerate to within the accuracy of the calculation.

In considering the symmetries involved in the ro-vibrational wave function it is necessary to consider both the permutation symmetry of the atoms in the molecule and the symmetry of the rotational portion of the wave function. The group theory involved in this problem has been described by Bunker[81] and Ezra.[82]

For the triatomics A_3, AB_2, and ABC, permutation symmetries S_3, S_2, and S_1 apply, respectively. In our coordinate system it is not possible to adapt the wave function a priori to S_3 symmetry.[30] S_1 has no permutation symmetry; so we will confine ourselves to a consideration of AB_2 systems.

In the flexible coordinates of Fig. 1, the interchange symmetry can be carried by the angular or the radial coordinates depending on the choice of (g_1, g_2). In scattering coordinates, $(g_1 = 0.5, g_2 = 0.0)$ for an atom–diatom system with a homonuclear diatomic, interchanging the like atoms is equivalent to changing $\theta \to \pi - \theta$. In a FBR, this symmetry is naturally carried by Legendre polynomials which have the property that even polynomials are symmetric (denoted $q = 0$) with respect to interchange of the two like atoms, and odd polynomials are antisymmetric ($q = 1$).

If coordinates are chosen in which $g_1 = g_2$ (these include both Radau and bond length–bond angle coordinates provided that the odd atom is chosen as the central atom), then interchanging the atoms is equivalent to the exchange $r_1 \leftrightarrow r_2$. Symmetrization in this case requires the selection of identical functions, $|m\rangle$, to carry the r_1 motions to those, $|n\rangle$, used for the r_2 coordinate. In this case symmetric functions can be written

$$|m, n, q = 0\rangle = 2^{-1/2}(1 + \delta_{mn})^{-1/2}(|m\rangle|n\rangle + |n\rangle|m\rangle), \quad m \geq n \tag{34}$$

and antisymmetric functions

$$|m, n, q = 1\rangle = 2^{-1/2}(|m\rangle|n\rangle - |n\rangle|m\rangle), \quad m > n \tag{35}$$

However, whereas the angular symmetrization is consistent with embedding the body-fixed z-axis along either r_1 or r_2 ($a = 0$ or 1), the radial symmetrization is only achieved with the z-axis embedded along the bisector of r_1 and r_2 ($a = 1/2$).[187] This means that for calculations involving rotational excited states our Hamiltonian cannot be fully symmetrized for coordinates with $g_1 = g_2$.

For each embedding the rotational functions are symmetrized by considering the behavior of the projection of the total angular momentum on the body-fixed z-axis, k, for the combined rotational and bending functions. For embedding $a = 0$ or 1, the symmetrized functions take the form

$$\begin{aligned}|j, k, p\rangle = 2^{-1/2}(1 + \delta_{k0})^{-1/2}[\Theta_{jk}(\theta)D^J_{Mk}(\alpha, \beta, \gamma) \\ + (-1)^p\Theta_{j-k}(\theta)D^J_{M-k}(\alpha, \beta, \gamma)], \quad k \geq 0\end{aligned} \tag{36}$$

The total parity is given by $(-1)^{J+p}$ with $p = 0$ or 1. States with $p = 0$ and 1 are conventionally labeled e and f states, respectively.[83]

This rotational symmetrization does not affect the effective vibrational kinetic energy operators, $\hat{K}_V^{(1)}$ and $\hat{K}_V^{(2)}$, or the potential energy term given above. It modifies the $\hat{K}_{VR}$ terms by introducing a factor of $\sqrt{2}$ into terms coupling $p = 0$, $k = 0$ to $p = 0$, $k = 1$. Of course, in this symmetrized basis, basis functions differing in p are decoupled. Furthermore, for each block it is necessary only to consider k values running from p to J.

The observation that the $p = 1$ secular matrix is simply a submatrix of the $p = 0$ secular matrix (with the rows and columns involving $k = 0$ removed) means that this matrix need not be explicitly calculated. In our programs, the $p = 0$ secular matrix is stored on disk and the appropriate submatrix is read and diagonalized for the $p = 1$ calculation.[39,84]

The inclusion of permutation symmetry in DVR space is not quite as straightforward as when working with a FBR. One procedure, recently proposed by Whitnell and Light,[85] is to develop separate symmetrized basis sets which can then be transformed to a DVR. This approach has the disadvantage that these basis functions are not in general the same as the basis functions of choice for the problem in question.

An alternative procedure is possible if one first transforms to the DVR and identifies symmetry related points in coordinate space. For example, in the AB_2 systems represented in scattering coordinates and discussed above, the symmetry is carried by associated Legendre polynomials. Functions with $j + k$ even are symmetric and those with $j + k$ odd antisymmetric with respect to reflection about $\theta = 90°$. In this case the symmetry of the Gauss-associated Legendre quadrature about $\chi = \cos\theta = 0$ means that all the unique problems lie in the half range $0 \leq \chi_{k\alpha} < 1$. It is then possible to symmetrize the DVR wave function $\phi_{\alpha t}^{Jk}$ by writing

$$\phi_{\alpha t}^{Jkq}(r_1, r_2; \chi_{k\alpha}) = 2^{-1/2}[\phi_{\alpha t}^{Jk}(r_1, r_2; \chi_{k\alpha}) + (-1)^q \phi_{\alpha t}^{Jk}(r_1, r_2; -\chi_{k\alpha})]; \qquad q = 0, 1, \quad \chi_{k\alpha} > 0 \tag{37}$$

where we have restricted ourselves to an even number of DVR points, N, to avoid the special case of $\chi_{k\alpha} = 0$.

Transforming the effective Hamiltonian to this symmetrized basis leaves its form unchanged but redefines the coupling matrices

$$L_{\alpha'\alpha}^{kq} = 2 \sum_{l=k}^{N/2+k-1} T_{2l+q,\alpha'}^{k}(2l+q)(2l+q+1)T_{2l+q,\alpha}^{k} \tag{38}$$

$$Q_{\alpha'\alpha}^{kq\pm} = 2 \sum_{l} T_{2l+q,\alpha'}^{k\pm 1} C_{2l+q,k}^{\pm} T_{2l+q,\alpha}^{k} \tag{39}$$

where the sum over l runs from the maximum of $(k, k \pm 1)$ to the minimum of $(N/2 + k - 1,\ N'/2 + k \pm 1 - 1)$. Thus, in the symmetrized DVR, only the terms that differ between even ($q = 0$) and odd ($q = 1$) calculations are provided by the $\mathbf{L}^{kq}$ and $\mathbf{Q}^{kq\pm}$ matrices.

This symmetrization scheme has considerable computational advantages. First, the frozen α, two-dimensional problems are independent of these matrices. Second, the bulk of the computer time in a DVR calculation is actually spent in transforming radial matrix elements with solutions of the two-dimensional problem; see equation (68). These transformed matrix elements, $\tilde{\mathbf{W}}$, and the eigenvalues of the two-dimensional problems, $\boldsymbol{\varepsilon}^k$, can easily be saved and used to generate solutions of both symmetries for little more cost than performing a calculation with only one symmetry. Full details of this symmetrization procedure have been given by Tennyson and Henderson.(31)

4. Rotational Excitation

4.1. Two-Step Procedure

The previous section discussed methods for calculating vibrational wave functions of triatomic systems. In fact, the formalism is presented in a sufficiently general manner that it could be used directly to perform fully ro-vibrational calculations. Indeed, until 1986, this was how calculations that included full ro-vibrational or Coriolis coupling were performed.

Analysis of the angular basis functions given by equation (10) shows that for a direct solution of the ro-vibrational problem the size of the secular problem increases as $(2J + 1)$. Symmetrization reduces this to two separate secular matrices increasing as J and $J + 1$, but it does not solve the basic difficulty with large J calculations. The result of this is that this approach has never been used for calculations with J larger than 4.

A way to avoid this problem, using a two-step variational procedure, was suggested by Tennyson and Sutcliffe.(50) The first step of this procedure is to solve the "vibrational" problem obtained by ignoring the off-diagonal Coriolis coupling terms. This approximation is equivalent to assuming k, the projection of the total angular momentum on the body-fixed z-axis, is a good quantum number. For certain systems this approximation is very accurate.(49,86,87)

Ignoring rotational symmetry,* the eigenfunctions of the effective, Coriolis decoupled Hamiltonian

$$\hat{H}^k(r_1, r_2) = \hat{K}_V^{(1)} + \hat{K}_V^{(2)} + \delta_{k'k}(\hat{K}_{VR}^{(1)} + \hat{K}_{VR}^{(2)}) + \langle j', k|V(r_1, r_2, \theta)|j, k\rangle_\theta \tag{40}$$

* Only unsymmetrized functions are used here because the rotational symmetrization is performed as part of the second step of the calculation.

can be written

$$|i, k\rangle = \sum_{jmn} c_{jmn}^{Jki} |j, k\rangle |m\rangle |n\rangle \tag{41}$$

with corresponding eigenenergy ε_i^{Jk}. The second step of the procedure then consists of using these eigenfunctions, symmetrized, as a basis for the exact effective Hamiltonian (11). This gives a Hamiltonian matrix of the form

$$\langle i', k'|\hat{H}|i, k\rangle = \delta_{k'k}\, \delta_{i'i}\varepsilon_i^{Jk} + \delta_{k'k\pm 1}(1 + \delta_{k0} + \delta_{k'0})^{1/2}\langle i', k'|\hat{K}_{VR}^{(1)} + \hat{K}_{VR}^{(2)}|i, k\rangle \tag{42}$$

where symmetrizing the rotational basis set, according to equation (36), introduces a factor of $\sqrt{2}$ in off-diagonal terms involving $k = 0$.

The form of this Hamiltonian is further simplified in orthogonal coordinates where $\hat{K}_{VR}^{(2)} = 0$ and the off-diagonal element is given by

$$\begin{aligned}\langle i', k \pm 1|\hat{K}_{VR}^{(1)}|i, k\rangle &= -\left\langle i', k \pm 1 \left| \delta_{j'j}\, \delta_{n'n} C_{Jk}^{\pm} C_{jk}^{\pm} \frac{\hbar^2}{2\mu_1 r_1^2} \right| i, k \right\rangle \\ &= -C_{Jk}^{\pm} \sum_{jmn} C_{jk}^{\pm} c_{jmn}^{Jki} \sum_{n'} c_{jmn'}^{Jk\pm 1 i'} \left\langle n' \left| \frac{\hbar^2}{2\mu_1 r_1^2} \right| n \right\rangle \end{aligned} \tag{43}$$

for the $a = 0$ embedding. The analogous expression for the $a = 1$ embedding is obtained by switching $r_1 \leftrightarrow r_2$, $\mu_1 \leftrightarrow \mu_2$ and quantum numbers $m \leftrightarrow n$.

Solving the rotational problem in this fashion has major advantages. First, it is not necessary to include all the solutions of the first step to obtain converged solutions for the second. The best algorithm for this[88] is to select the intermediate basis functions according to an energy ordering criteria, that is, according to the ε_i^{Jk}. This results in a greatly reduced final secular matrix. This reduction in size is particularly drastic for the case where k is a nearly good quantum number; thus the importance of selecting the best embedding.

Second, the secular matrix constructed for the second step has a characteristic sparse structure; see Fig. 2. All the elements are zero with the exception of the diagonal elements and one off-diagonal block linking k with $k \pm 1$. Thus, it is only necessary to store the nonzero elements, reducing the core requirement by a factor of approximately J.[89] Furthermore, this sparse matrix can rapidly be diagonalized using an iterative technique to obtain the eigenenergies and wave functions of interest (see Section 6.5). This diagonalization is so efficient that it is actually quicker to solve the full problem, the one obtained by not truncating the intermediate basis, in two steps than directly in one step.[89]

Finally, within the symmetrized rotational basis, the secular matrix for $p = 1$ is simply a submatrix of that for $p = 0$. The solution to both problems can thus be obtained using the same matrix by simply removing the portion of the $p = 0$ matrix involving $k = 0$.[64]

For this symmetrized problem, the sth solution of the second step can be written

$$\Psi_s^{Jp} = \sum_{ik} b_{ik}^{Jps} |i, k, p\rangle \tag{44}$$

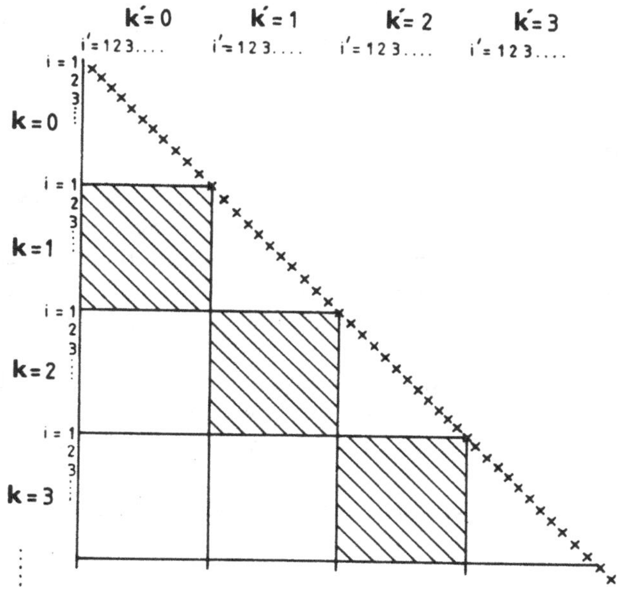

Figure 2. Structure of the secular matrix for the second variational step.

with corresponding eigenenergy E_s^{Jp}. The coefficient vectors $\mathbf{b}^{Jps}$ can then be back transformed to yield coefficients of the wave function in terms of the original basis functions

$$d_{kjmn}^{Jps} = \sum_i b_{ik}^{Jps} c_{jmn}^{Jki} \tag{45}$$

These coefficients are important for calculating properties of the wave function and in particular transition intensities that are discussed below.

There is one important difference between the two-step procedure outlined above and that proposed by Chen et al.[22] Whereas our procedure solves a new "vibrational" problem for the $(J, |k|)$ combination, that of Chen et al. simply uses the solutions of the $J = 0$ problem to expand all rotationally excited problems of interest. This obviously leads to a substantial saving in the number of vibrational problems that have to be solved. The problem with Chen et al.'s method, which has also been used by Bunker, Jensen, and co-workers with their MORBID Hamiltonian (see, for example, Refs. 90, 91), is that the intermediate basis functions do not allow for rotational distortion effects and, more seriously, the method fails for bent molecules when they start sampling linear geometries. This means that for studies involving any large-amplitude vibrations or extreme rotational motion, the method of Tennyson and Sutcliffe needs to be employed. Other workers to adapt this method to their formalism include Carter and Handy,[92] and Choi and Light.[46]

4.2. Adaptation for the DVR

There are two possible ways to adapt the two-step method outlined above to the discrete variable formalism discussed in Section 3.5. Adaptation to the first step is straightforward because neglecting the Coriolis coupling simply removes the term involving $\mathbf{Q}^{k\pm}$ from equation (29). This also means that each k-fixed DVR calculation relies only on a single Gauss-associated Legendre quadrature scheme.

Tennyson and Henderson[31] chose to approach the second step of this problem by back transforming the solution vectors, $\mathbf{b}^{Jki}$, of the k-fixed DVR problem to the solution vectors of the equivalent FBR basis:

$$c_{jmn}^{Jki} = \sum_\alpha T_{\alpha j}^k b_{mn\alpha}^{Jki} = \sum_\alpha \omega_{k\alpha}^{1/2} \Theta_{jk}(\theta_{k\alpha}) b_{mn\alpha}^{Jki} \tag{46}$$

This had the advantage that the program ROTLEVD[39] could be used without alteration for the second step in the variational calculation.

An alternative procedure is not to transform the solutions but to rewrite the second step in a DVR. In this case

$$\langle i', k\pm 1|\hat{K}^{(1)}_{VR}|i, k\rangle = -\left\langle i', k\pm 1\left|\delta_{n'n}C^{\pm}_{Jk}Q^{k\pm}_{a'a}\frac{\hbar^2}{2\mu_1 r_1^2}\right|i, k\right\rangle$$

$$= -C^{\pm}_{Jk}\sum_{mna} b^{Jki}_{mna}\sum_{n'a'} b^{Jk\pm 1i'}_{mn'a'}Q^{k\pm}_{a'a}\left\langle n'\left|\frac{\hbar^2}{2\mu_1 r_1^2}\right|n\right\rangle \quad (47)$$

It can be seen that the presence of the extra summation (over α') compared to equation (43) suggests that this approach, though more direct, will be less efficient than back transforming.

It should be noted that other considerations will come into play with a calculation that is performed entirely within a DVR. In this case, invoking the quadrature approximations shows that the radial matrix element will be diagonal. Furthermore, experience with single DVR calculations suggests that the back transformation of three DVRs is likely to be cumbersome because of the large number of basis functions required to carry the more compact DVR. In this case the direct method is likely to become very much more attractive.

5. Simulated Spectra

5.1. Linestrengths

A natural extension of the calculation of ro-vibrational wave functions is to use these wave functions to study properties of the system as a function of ro-vibrational state. Perhaps the most important property of the system from a spectroscopic point of view is the transition intensity. It is this intensity that determines which transitions are seen. Conversely, given an observed intensity and a value for the linestrength, spectra can yield important physical information about the system studied. It is this that makes spectra so useful as probes in astrophysics.

In 1984, Brocks et al.[52] used ab initio dipole surfaces of the floppy molecules KCN and LiCN to compute intensities for a range of vibrational transitions. They used this data to synthesize stick spectra, the main prediction of which concerned the complexity of the vibrational spectra of these systems at temperatures at which they were usually prepared experimentally. A prediction supported by the great difficulty in making any assignments in the observed infrared spectra.[93]

Brocks et al.'s theory only covered $J = 1 \leftarrow 0$ transitions and was applied to wave functions generated neglecting off-diagonal Coriolis interactions. This approximation, later relaxed,[53] is not reliable for transition intensities, even for systems where it yields good transition frequencies, because a small amount of mixing can greatly enhance an otherwise weak transition.

Recently, Miller et al.[56] have completely generalized the theory of Brocks et al. Using the FBR wave functions outlined above, they obtained an expression for the linestrength

$$\begin{aligned} S(u - l) = {} & \tfrac{1}{4}[(2J' + 1)(2J'' + 1)] \\ & \times \Bigg\{ \sum_{\nu=-1}^{+1} \sum_{\lambda=|\nu|} \sum_{k=p''}^{J''} \sum_{j''j'} a(\nu, \nu + k, \lambda)[(2j' + 1)(2j'' + 1)]^{1/2} \\ & \times \begin{pmatrix} J' & 1 & j'' \\ -k-\nu & \nu & k \end{pmatrix} \begin{pmatrix} j' & \lambda & j'' \\ 0 & 0 & 0 \end{pmatrix} \begin{pmatrix} j' & \lambda & j'' \\ -k-\nu & \nu & k \end{pmatrix} \\ & \times \sum_{m''n''} \sum_{m'n'} B_{\lambda,\nu}^{m'm''n'n''} \times d_{k'j'm'n'}^{J'p's'} d_{k''j''m''n''}^{J''p''s''} [(-1)^{J''+J'+1} + (-1)^{p''+p'}] \Bigg\}^2 \end{aligned} \tag{48}$$

where the primes and double primes refer to the upper (u) and lower (l) states, respectively, and the vectors $\mathbf{d}^{Jps}$ are coefficients of these states in the basis set expansion of the wave function. The extra angular factor in equation (48) is given by

$$\begin{aligned} a(0, n, \lambda) &= 1, \qquad 0 \le n \le J' \\ a(\pm 1, 0, \lambda) &= -[\lambda(\lambda + 1)]^{1/2} \\ a(\pm 1, n, \lambda) &= -\left[\frac{\lambda(\lambda + 1)}{2}\right]^{1/2}, \qquad 0 < n \le J' \\ a(\nu, n, \lambda) &= 0, \qquad n < 0, \quad n > J', \quad \nu = 0, \pm 1 \end{aligned} \tag{49}$$

In equation (48), the initial sum runs over all the components of the dipole in the body-fixed axis systems. In this expression, the dipole components, μ_ν, have been expanded in terms of (associated) Legendre polynomials in an analogous fashion to the potential expansion (20)

$$\mu_\nu^m(r_1, r_2, \theta) = \sum_{\lambda=0} \left(\frac{2}{2\lambda+1}\right)^{1/2} B_{\lambda,0}(r_1, r_2)\Theta_{\lambda,0}(\theta), \qquad \nu = 0$$
$$= \sum_{\lambda=1} \left(\frac{1}{2\lambda+1}\right)^{1/2} [\lambda(\lambda+1)]^{1/2} B_{\lambda,\nu}(r_1, r_2)\Theta_{\lambda,\nu}(\theta), \qquad \nu = \pm 1 \tag{50}$$

with $B_{\lambda,-1}(r_1, r_2) = B_{\lambda,+1}(r_1, r_2)$. Again, this expansion can be computed numerically for each (r_1, r_2) of interest, using the appropriate Gaussian quadrature scheme. The matrix elements $B_{\lambda,\nu}^{m'm''n'n''}$ are then obtained by integrating over the appropriate radial basis functions.

The sum rules applying to the $3-j$ symbols in equation (48) and the phase factors at the end of the equation impose the selection rules for all triatomic systems:

$$\begin{aligned} \Delta J &= 0, \qquad \Delta p = \pm 1 \\ \Delta J &= \pm 1, \qquad \Delta p = 0 \end{aligned} \tag{51}$$

Additional selection rules that depend on specific molecular symmetries occur as a result of the structure of the wave function computed from the basis function. For example, AB_2 systems have the additional selection rule

$$\Delta q = 0 \tag{52}$$

The only selection rules obtained by this formalism are rigorous and arise purely from symmetry considerations, either rotational or permutational. This means that any propensity rules, based for example on the harmonic oscillator model, will only be seen in the calculation to the same extent as they are observed experimentally. This has the advantage that spectra may be computed without any a priori assumptions about which transitions are important. A number of predictions of conventionally "forbidden" spectra have already been made.[94,95]

The Einstein A_{ul} coefficient for spontaneous emission from state u to a lower state l is given by*

$$A_{ul} = \frac{1}{(2J'+1)} \frac{64\pi^4\omega_{ul}^3}{3h} S(u-l) \tag{53}$$

* Note that this expression is in error by a factor of π^2 in Ref. 56.

where the frequency of the transition is given by $\omega_{ul} = E_u - E_l$. This form is particularly important for astrophysical application.[58,94,96] With the Einstein A_{ul} coefficients it is also possible to calculate the fluorescent lifetime, τ_u, of a particular state[53,97]

$$\tau_u = \frac{1}{\sum_{l<u} A_{ul}} \tag{54}$$

5.2. Temperature Dependence

To obtain absolute transition intensities it is necessary to account for the population of individual levels. Values for the integrated absorption coefficient may be obtained from[98]

$$I(\omega_{ul}) = \frac{8\pi^3 \omega_{ul} g_l [\exp(-E_l/kT) - \exp(-E_u/kT)]}{3hQc} S(u-l) \tag{55}$$

where g_l is the spin-plus-symmetry degeneracy of the lower state which has energy E_l; h is Planck's constant; k is Boltzmann's constant; T is the temperature; and the partition function, Q, is given by

$$Q = \sum_i g_i(2J_i + 1) \exp\left(-\frac{E_i}{kT}\right) \tag{56}$$

The temperature dependence of a spectrum thus manifests itself via the partition function and the Boltzmann factors. Calculation of the partition function is in principal straightforward. However, it may well require the determination, at least approximately, of energies for states which are otherwise not involved in the spectral region of interest.

Nuclear spin statistics, which are a consequence of the Pauli principle, enter the formulae as a simple multiplicative factor which is a constant for all states with a given permutation symmetry. Adding this factor is easy provided the ro-vibrational calculation is performed using the full permutation symmetry. If lower symmetry is used, then it is necessary to assign symmetry labels to each state before the partition function can be correctly determined. For example, H_3^+ has a S_3 permutation symmetry but our calculations[30,31,95,96,99–101] have all been performed using only the S_2 permutation group. In this representation, states of A_1 and A_2 symmetry have even and odd symmetry, respectively, and the degenerate E states are split between the two S_2 representations. The nuclear spin statistics mean that the spin-plus-symmetry factor g_i is 0, 4, and 2 for A_1, A_2, and E states, respectively.

The partition function can thus be calculated using the odd S_2 levels, but an accurate determination requires each of these levels to be labeled as either A_2 or E. Even for moderate temperature this involves labeling many hundreds of levels by hand.

The velocity of the individual molecules in a given experiment is related to the temperature. The Doppler width of a given transition will therefore also be temperature dependent. This is considered below.

5.3. Line Profiles

With a knowledge of the linestrength of the transitions of interest and the partition function, it is possible to synthesize stick spectra if one assumes a Boltzmann distribution for the system. These stick spectra, examples of which are shown in Figs. 3 and 7, can be very useful for interpreting observed spectra but are inadequate for very congested spectral regions. In these regions it may be necessary to consider several lines together. To do this one requires some knowledge of the relevant line profiles.

There are four things that need to be considered when determining the line profile of an observed transition. Each line has an intrinsic width, given by the uncertainty principle, due to the finite lifetime of the states involved in the transition. However, except for quasi-bound levels, which are beyond the scope of this chapter, this intrinsic width is usually too narrow to be

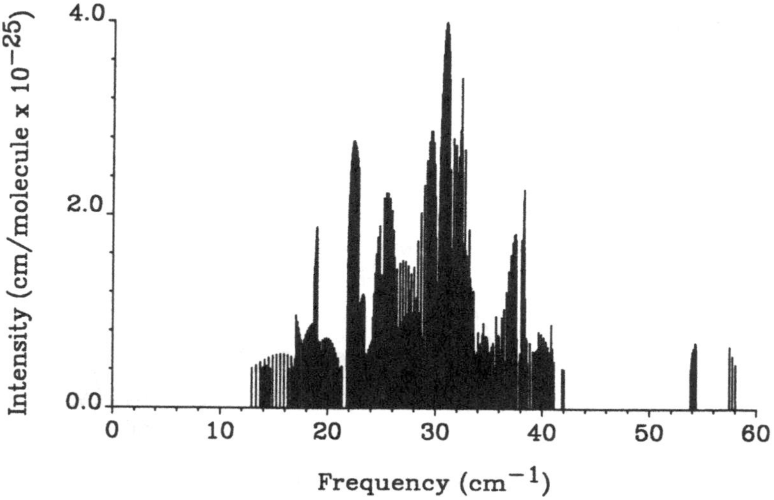

Figure 3. Synthesized stick spectrum for the Ar–N_2 Van der Waals complex (temperature = 87 K).[102]

significant. Collisional effects can also cause broadening of spectral lines; this effect is clearly pressure dependent. A particular transition will have a Doppler width. This width may be temperature dependent, but many experiments are now constructed so that the spectra obtained are Doppler-free. Finally, instrumental considerations in any experiment yield a finite resolution or machine factor for a given transition. The appropriate general profile is thus a convolution of a Lorentzian, generally used to represent lifetime and collision effects, and a Gaussian which is used to represent Doppler broadening. For many experiments the machine factor is also approximately a Gaussian.

Figures 3 and 4 give a simple illustration of the importance of using line profiles rather than stick spectra when synthesizing spectra for highly congested regions. These spectra come from the same, recent calculation on the N_2Ar Van der Waals complex.[102] These calculations were designed to explain the structure in McKellar's[103] observed N_2Ar spectra which were obtained at liquid nitrogen temperatures.

The calculations determined the intensity of all possible transitions between bound states of the complex—more than 30,000 of them. The 15,000 or so transitions with intensity greater than 0.1% of the most intense transition were then used to synthesize spectra. The resulting stick spectrum, Fig. 3, is very congested but does suggest some structure. When the lines in this spectrum are given an appropriate shape (the major effect was assumed to

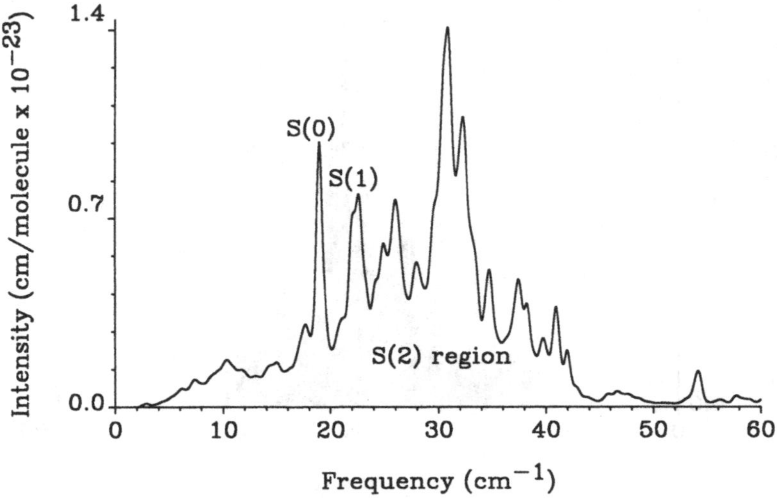

Figure 4. Synthesized spectrum with Gaussian line profiles for the $Ar–N_2$ Van der Waals complex (temperature = 87 K; total density = 4.4 amagat).[102]

be pressure broadening), the result is Fig. 4. It is noteworthy that convoluting the lines results in a spectrum that is about 30 times more intense than that suggested by the stick spectrum. Furthermore, there is also a shift in the relative importance of various spectral features between the two spectra.

6. Computational Considerations

6.1. Radial Matrix Elements

All the angular matrix elements required to solve ro-vibrational problems were obtained analytically in Section 3 and used to obtain the effective radial Hamiltonian of equation (11). So far we have suggested appropriate radial basis functions for this Hamiltonian, but no mention has been made of how the radial matrix elements are actually calculated.

In general our philosophy is: for each coordinate, use orthogonal polynomials as basis functions, solve as many of the integrals as possible analytically, and then use a Gaussian quadrature scheme based on the same polynomials. This approach can be seen as a natural generalization of the work of Whitehead and Handy[3,4] who solved Watson's Hamiltonian using Hermite polynomials to carry the vibrational motions and Gauss–Hermite quadrature for integrals such as those over the potential and the Watson term. A natural consequence of this philosophy, which forswears approximations or potentials in specific coordinates, is that it is possible to obtain effectively exact solutions given only a potential and the Born–Oppenheimer approximation.[57]

For the Morse oscillatorlike functions detailed in Section 3.3, all the integrals over the derivative kinetic energy operators can be performed analytically[14]

$$\left\langle n\left|\frac{\partial^2}{\partial r^2}\right|n'\right\rangle = \frac{\beta^2}{4}\{\delta_{n'n}[2n(\alpha+n+1)+\alpha+1] - \delta_{n'\pm 2,n}[(\alpha+n+1\pm 1) \times (\alpha+n\pm 1)(n+1\pm 1)(n\pm 1)]^{1/2}\} \tag{57}$$

and[17]

$$\left\langle n\left|\frac{\partial}{\partial r}\right|n'\right\rangle = \frac{\beta}{2}\delta_{n',n\pm 1}\left[\left(n-\frac{3}{2}\pm\frac{1}{2}\right)\left(\alpha+n-\frac{3}{2}\pm\frac{1}{2}\right)\right]^{1/2} \tag{58}$$

However, the integrals over the inverse radial operators r^{-2} and r^{-1}, as well of course as over the potential, have to be performed numerically.

Spherical oscillator functions have so far only been used with orthogonal coordinate systems.[17,30,31] In this case all the kinetic energy matrix elements can be evaluated analytically[17]

$$\left\langle n \left| \frac{\partial^2}{\partial r^2} \right| n' \right\rangle = \frac{\beta^{1/2}}{2} \left[\delta_{n'n}\left(2n + \alpha + \frac{3}{2}\right) + \delta_{n'\pm 1,n}\left(\frac{n + \frac{1}{2} \mp \frac{1}{2}}{n + \alpha + 1 \mp \frac{1}{2}}\right)^{1/2} \right] + \langle n|r^{-2}|n'\rangle \tag{59}$$

$$\langle n|r^{-2}|n'\rangle = \frac{\beta^{1/2}}{2}\left[\frac{n!}{n'!}\,\frac{\Gamma(n' + \alpha + \frac{3}{2})}{\Gamma(n + \alpha + \frac{3}{2})}\right]^{1/2} \times \sum_{\sigma=0}^{n'} \frac{n'!}{\sigma!}\,\frac{\Gamma(\sigma + \alpha + \frac{1}{2})}{\Gamma(n' + \alpha + \frac{3}{2})}, \qquad n' \le n \tag{60}$$

where the numerical reason for the apparently inefficient grouping of the factorial and Γ-function terms in equation (60) will be explained in the following section.

6.2. Gaussian Quadrature Schemes

One reason for the particular choice of Morse oscillatorlike functions by Tennyson and Sutcliffe,[14] besides the completeness problem already discussed, is that it is very much easier to work with a set of Laguerre polynomials of the same degree, α, rather than one where the degree is a function of energy level as with the standard Morse oscillators. This means that with a Gaussian quadrature scheme based on Laguerre polynomials of the same degree, overlap integrals can be evaluated exactly and the integral over any well-behaved operator can be calculated cheaply and efficiently.[69]

A general program for the calculation of Gauss–Laguerre quadrature points has been given by Stroud and Secrest.[68] However there are numerical problems with the weights in Gauss–Laguerre quadrature schemes. For even moderate values of α the weights are liable to cause underflow problems on most computers. Stroud and Secrest warn against using their program for $\alpha > 30$. However, for the Morse oscillatorlike functions, the value of α is related* to the number of bound states in that coordinate. For strongly bound systems there may be several hundred bound stretching states associated with a particular coordinate.

* In practice this relationship is only loose because the parameters that give α are usually variationally optimized.

Tennyson and Sutcliffe[14] found a way to avoid this problem. For an S-point quadrature scheme, they transferred a factor of $\Gamma(S+\alpha+1)$ between the weights and the normalization constant $N_{n\alpha}$. Thus, the new weights, w_i', and normalization constants, $N'_{n\alpha}$, are given by

$$w_i' = s!\left[\frac{dL_S^\alpha(x_i)}{dx} L_{S-1}^\alpha(x_i)\right]^{-1} = \frac{w_i}{\Gamma(S+\alpha+1)} \tag{61}$$

$$N'_{n\alpha} = \left[n!\,\frac{\Gamma(S+\alpha+1)}{\Gamma(n+\alpha+1)}\right]^{1/2} = N_{n\alpha}[\Gamma(S+\alpha+1)]^{1/2} \tag{62}$$

where x_i is the S-point Gauss–Laguerre quadrature point that has weight w_i. This trick places the terms in $\Gamma(\alpha)$ as ratios which can then be evaluated as binomial coefficients. The result is numerically stable for all systems on computers with a reasonable exponent range.

The resulting approximation to the radial integral over the potential energy function can then be written

$$\begin{aligned}\langle m, n|V_\lambda(r_1, r_2)|m', n'\rangle = {} & N'_{m\alpha_1}N'_{m'\alpha_1}N'_{n\alpha_2}N'_{n'\alpha_2}f_1(\beta_1)f_2(\beta_2)\\ & + \sum_{i=1}^{S}\sum_{j=1}^{T} w_i'w_j'L_m^{\alpha_1}(x_i)\\ & \times L_{m'}^{\alpha_1}(x_i)L_n^{\alpha_2}(x_j)L_{n'}^{\alpha_2}(x_j)V_\lambda\big(r_1(x_i), r_2(x_j)\big)\end{aligned} \tag{63}$$

where the points x_i and x_j are the zeros of $L_S^{\alpha_1}$ and $L_T^{\alpha_2}$, and related respectively to r_1 and r_2 through equations (22) and (25) for Morse oscillatorlike functions and spherical oscillators, respectively. The additional normalization functions, $f(\beta)$, equal β for Morse oscillatorlike functions and $2\sqrt{\beta}$ for the spherical oscillators. For well-behaved potential functions, exact integrals can be obtained in this fashion if $S > m + m'$ and $T > n + n'$.[69]

The inverse radial operators between Morse oscillatorlike functions can be evaluated using

$$\langle n|r^{-j}|n'\rangle = N'_{n\alpha}N'_{n'\alpha}\beta \sum_{i=1}^{S} w_i'L_n^\alpha(x_i)L_{n'}^\alpha(x_i)\left[r_e + \beta^{-1}\ln\left(\frac{A}{x_i}\right)\right]^{-j}, \qquad j = 1, 2 \tag{64}$$

6.3. Basis Set Selection

Performing a calculation within a finite basis representation (FBR) gives some flexibility about how this basis set is actually chosen. Usually for triatomics the internal coordinate basis is simply a product of one-dimensional basis functions for each coordinate. If one then simply takes all terms in each expansion up to some N_i ($i = 1, 2, \theta$ for r_1, r_2, θ, respectively), then one gets a basis set of dimension $(N_1+1)(N_2+1)(N_\theta+1)$. This basis contains the function (N_1, N_2, N_θ) but not $(N_1+1, 0, 0)$, $(0, N_2+1, 0)$ or $(0, 0, N_\theta+1)$. It is therefore likely that this is not the best method of choosing a product function.

Two methods of preselecting basis functions for FBR calculations have been tried by us. One method calculates the diagonal matrix element for a large number of candidate basis functions. The final basis is then chosen as the functions that have the N lowest diagonal elements, where N is the size of secular matrix desired. This approach can be regarded as loosely founded on perturbation theory. The logical next step has been taken by Hutson and Le Roy[104] and Hutson[105] who developed a method that used perturbation theory to give the contribution from functions omitted from the basis set expansions.

An alternative method is based on quantum numbers. All functions are selected that satisfy the relationship

$$Q \geq \frac{n_1}{d_1} + \frac{n_2}{d_2} + \frac{n_\theta}{d_\theta} \tag{65}$$

where n_i are the number of quanta in mode i. The d_i serve to weight this selection because one mode—for example, a low-frequency bend—may require more functions than some other mode. This method has also been used by other workers; see, for example, Refs. 22 and 106.

Both these methods, which can also be used in hybrid form, are both found to give greatly enhanced convergence compared to taking the full product. Further discussion, including comparison of convergence with the different selections criteria, can be found in Ref. 10.

Finally, it should be noted that for DVR-based methods, the intermediate diagonalization and truncation serve a similar purpose to basis set selection but in an altogether more rigorous fashion. We thus anticipate that this form of basis set preselection will not be an important feature of future DVR codes.

6.4. Construction of the Secular Matrix

Although the construction of the secular matrix is straightforward, given all the matrix elements, this step can become computationally expensive if not programmed carefully. In this section we will consider the construction of "vibrational" secular matrices associated with the effective FBR

Hamiltonian of equation (11) and the DVR Hamiltonian of equation (29) as well as the structured secular matrix of equation (43) obtained as the final step in the two-step calculation of rotationally excited states. All these secular matrices are real symmetric so only the lower triangle need actually be evaluated.

The conventional FBR secular matrix is best constructed by letting the loops over radial basis functions run faster than those over the angular functions. This is because the most time-consuming piece of the secular matrix construction is the evaluation of Gaunt coefficients; see equation (21). Indeed, our recent programs[39] construct this matrix a band at a time, the width of each band being the number of radial functions, $|m\rangle|n\rangle$, associated with the current angular quantum numbers, $|j, k\rangle$. After construction each band is written to disk. This is done for efficient core usage as the matrix $\langle m, n|V_\lambda|m', n'\rangle$ may sometimes be large and is beneficially overlaid with the secular matrix.

The DVR approach involves the construction and diagonalization of many secular matrices.[45] However, if, as discussed above, only one coordinate is described using a DVR, then it is only construction of the final secular matrix that is computationally expensive. For a given fixed value of k, the elements of this final matrix can be written[31]

$$\langle \alpha', t'|\hat{H}^k|\alpha, t|\alpha, t\rangle = \delta_{\alpha\alpha'}\,\delta_{tt'}\,\varepsilon^k_{\alpha t} + \tilde{W}_{\alpha' t' \alpha t} L^k_{\alpha'\alpha} \tag{66}$$

where $\varepsilon^k_{\alpha t}$ is the tth eigenenergy of the two-dimensional Hamiltonian $\hat{h}^{(2D)}(r_1, r_2)$, of equation (33), with corresponding wave function

$$|\alpha, t\rangle = \sum_{mn} a^{Jkt}_{mn\alpha}|m, n\rangle \tag{67}$$

In our work,[31,47] the radial basis functions have been composed of Morse oscillatorlike functions and/or spherical oscillator functions as in the FBR problems discussed above.

In equation (66) above, the off-diagonal matrix $\tilde{\mathbf{W}}$ is obtained by transforming the appropriate radial matrix elements. This transformation is greatly simplified by the use of orthogonal coordinates in which case it is

$$\tilde{W}_{\alpha' t' \alpha t} = \sum_{m,n}\sum_{m',n'} a^{Jkt}_{mn\alpha}a^{Jkt'}_{m'n'\alpha'}\left(\delta_{nn'}\left\langle m'\left|\frac{\hbar^2}{2\mu_1 r_1^2}\right|m\right\rangle + \delta_{mm'}\left\langle n'\left|\frac{\hbar^2}{2\mu_2 r_2^2}\right|n\right\rangle\right) \tag{68}$$

This transformation step is still found to dominate the CPU time requirement,[31] although it is possible to make savings by using the symmetry of the untransformed matrix elements with respect to both $m \leftrightarrow m'$ and $n \leftrightarrow n'$.

Construction of the final, Coriolis coupled secular matrix in the two-step variational procedure is also dominated by the need to transform matrix elements over the radial functions. In this case, for orthogonal coordinates the transformation, given by equation (43), is diagonal in all quantum numbers, except for one of the radial basis sets (which depends on the embedding). It is thus usually possible to construct this secular matrix faster than it is diagonalized. This is especially so because, as discussed in Section 4.1, it is possible to obtain results for both rotational symmetries by only constructing the $p = 0$ secular matrix.

6.5. Diagonalization

Diagonalization is a key step in all secular matrix methods. It is generally the computationally most expensive step, although we note that this may not be true for DVR-based methods. It is because of the expense both in storage and CPU time of diagonalizing matrices that great emphasis is placed in all methods in constructing efficient basis sets.

In our approach, there are four aspects to this. Basis sets are defined with adjustable parameters which can be optimized using the variational principle to give a compact representation. Basis sets may be preselected to retain only the most important product functions. The construction and diagonalization of intermediate problems, both in the DVR and two-step variational approaches, leads to very compact final basis sets. Finally, we always use orthonormal basis functions, in contrast to the use, for example, of distributed Gaussian functions by Bačić and Light,[42] because this simplifies the matrix diagonalization problem.

All the secular matrices constructed in the previous section are real symmetric. However, they have different characteristics which means that different diagonalization techniques are suited to each one.

The "vibrational" secular matrix obtained from the effective radial Hamiltonian in equation (11) generally has less than 10% of its eigenvalues which are physically significant. This problem is thus well suited to a diagonalizer designed to give only the lowest i solutions of an L-dimensional problem. Such methods are reckoned to be efficient when $i/L < 1/4$. We have found the EISPACK routine EIGSFM[107] very satisfactory for this problem, especially on Cray machines where a vectorized version is supplied in the Maths library. An out-of-core Lanczos diagonalization procedure has been used[28,108] for FBR problems but found to be very time-consuming.

The secular matrices obtained as the final step of the DVR procedure generally have many more significant eigenvalues, typically about 40%.[42] This means that a full diagonalization procedure is best for this problem. Such a procedure is also needed for the intermediate two-dimensional problems in the DVR method because many of the intermediate solutions are used as basis functions for the full problem. Of course, there are many real symmetric, full matrix diagonalization procedures available. We have found the NAG routine F02ABF[109] satisfactory for our purposes. We note that on vector processors, it is necessary to allocate space for the entire square matrix in order to vectorize either for EIGSFM or F02ABF.

The secular matrix for the final step in the two-step variational procedure has a sparse, blocked structure; see Fig. 2. Experimentation with diagonalizers for full real symmetric matrices, banded real symmetric matrices, and sparse matrices showed the sparse matrix diagonalizer to be more efficient in both storage and CPU time usage.[50,89] The algorithm selected is due to Nikolai[110] and is implemented as NAG routine F02FJF.[109] This method has the additional advantage, not considered in the original tests, that for our particular problem very efficient vectorization can be achieved. A detailed discussion of how this is done is given elsewhere.[64]

6.6. Transformations

After diagonalization, the next major user of computer time is transformations. These transformations arise from the use of intermediate representations, both in the DVR calculations and in the two-step variational approach. Each time an intermediate basis is employed, there are in principle two transformations that can arise: a forward transformation, which involves recasting matrix elements of the original representation in terms of the intermediate functions, and a back transformation, which takes the final coefficients of the wave function and expresses them as coefficients of the original basis set expansion or grid points.

The back transformation is the simpler of the two transformations. One can write the general coefficients of the ith eigenvector of the initial expansion as $c_\alpha^{\beta i}$. Here α represents the product of basis function quantum numbers and β the quantum numbers or grid points fixed at this stage of the calculation. The general coefficient of the sth expansion of the second step is then $b_{i\beta}^s$, where i runs over the first-step basis functions included in the second step. The transformation can then be written

$$d_{\alpha\beta}^s = \sum_i b_{i\beta}^s c_\alpha^{\beta i} \tag{69}$$

where $d_{\alpha\beta}^s$ is the coefficient of the sth state in terms of the now fully coupled original expansion.

This transformation, examples of which can be found in equations (45) and (46), is easily vectorized. The only problem may be managing all the vectors with limited storage capacity.

More computationally demanding is the transformation of matrix elements from the representation depending on $|\alpha\rangle$ to one depending on $|\beta, i\rangle$. In computational terms this operation can be written[64]

$$\tilde{w}_{i'i}^{\beta'\beta} = \tilde{w}_{i'i}^{\beta'\beta} + p_{i'}q_i \tag{70}$$

where

$$p_{i'} = c_{\alpha'}^{\beta'i'} w_{\alpha'\alpha}, \qquad q_i = c_{\alpha}^{\beta i} \tag{71}$$

and $w_{\alpha'\alpha}$ is a matrix element calculated in the original representation, which is assumed to be independent of β. Examples of this transformation are given by equations (43) and (68).

The operation represented by equation (70) is known as a rank-one update. Although the operation is implicitly vectorizable, experience has shown that CONVEX machines, which provide explicit rank-one update routines in the machine library, perform relatively better than Cray supercomputers for which, at least until very recently, no explicit rank-one update routines existed.

This particular transformation has strong similarities with the 4-index transformation common in molecular electronic structure calculations. Indeed, if α represents n indices [$n = 3$ for equation (43)], then equation (70) corresponds to the inner loop of a $2n$-index transformation. However, in the cases considered here, the transformations are simplified by the fact that all matrix elements that need to be transformed are diagonal in all except one of the n indices.

6.7. Program Structure

We have described above a number of different procedures that must be gone through in order to compute a ro-vibrational spectrum. Our method has been to implement each of these procedures as a separate program module and allow them to communicate via disk files. Figure 5 shows the structure of FBR codes (the DVR codes are currently less well developed, but it is to be anticipated that they have a similar general structure).

The main driver routine for all our FBR calculations is TRIATOM, for which basis functions may be preselected by SELECT. TRIATOM performs the vibrational calculations detailed in Section 3. It then provides data for ROTLEVD to perform the fully coupled rotational calculations when $J > 0$.

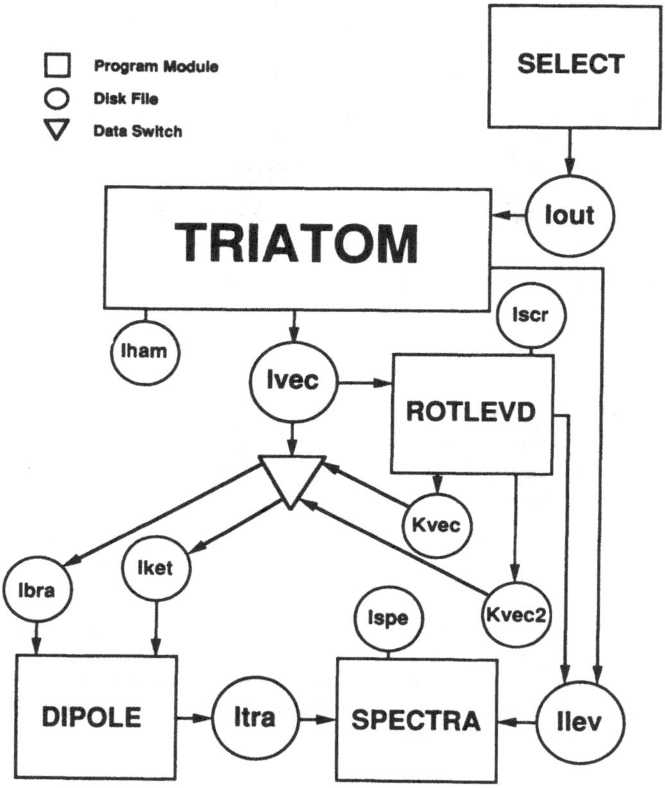

Figure 5. Program module flow diagram for the TRIATOM program suite.[39]

Both TRIATOM and ROTLEVD provide wave function data for the calculation of transition intensities and energy levels for transition frequencies, partition functions, and so on. The wave function files (containing appropriate energy levels) are stored separately for each set of good quantum numbers of the system, that is, (J, p) and any symmetry label, q, that may be appropriate. The energy levels are also stacked on a single file. These files can be used to study other properties of a particular system. For example, the wave function file is used to give contour plots which are important in the assignment of vibrational quanta. See Figs. 6 and 8 for examples.

Program DIPOLE processes pairs of wave function files to give transition intensities for all or some of the states contained in the bra and ket files. The transitions are stacked on a single transition file. This master file is then processed by SPECTRA which chooses a frequency range, physical

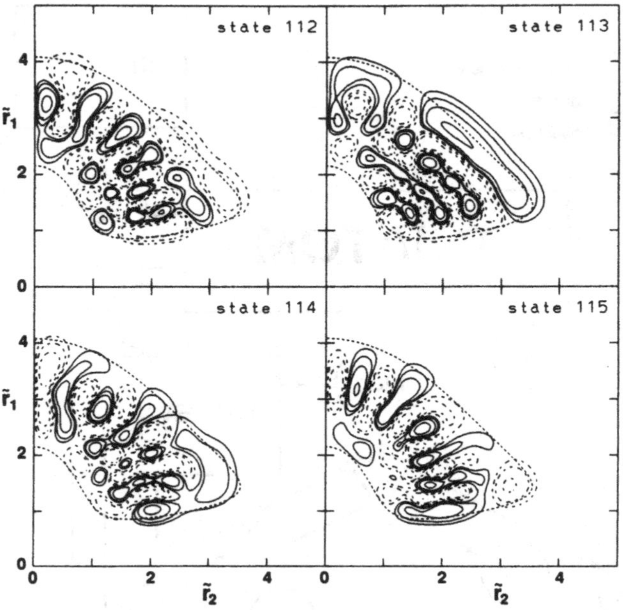

Figure 6. Contour plots of four high-lying H_3^+ $J = 0$ states.[31] The plots are $\tilde{r}_1$ versus $\tilde{r}_2$ in mass-scaled scattering coordinates with θ frozen at 90°. The weighting is $\tilde{r}_1 = \alpha r_1$ and $\tilde{r}_2 = r_2/\alpha$ where $\alpha = (3/4)^{1/4}$. Solid (dashed) contours enclose regions where the wave function has positive (negative) amplitude. Contours are drawn at 8%, 16%, 32%, and 64% of the maximum amplitude of the wave function.

conditions (in particular, temperature), and in our most recent versions, line profiles. The resulting output files(s) can be used directly to give graphical representations of the spectra, as can be seen in Figs. 3, 4, and 6.

The whole suite of programs is fully documented and has been published.[39]

7. Applications

7.1. H_3^+ and Its Isotopomers

The H_3^+ family, that is, H_3^+ itself, H_2D^+, D_2H^+, and D_3^+, has proved a very fruitful area of study for us.[30,31,50,51,56,78,94–96,99–101,111–114] The bound (and quasi-bound[115,116])-state dynamics of these systems are very rich. This feature, combined with accurate ab initio potential energy surfaces, including one of spectroscopic accuracy by Meyer et al.,[117] has led to a strong interaction between first principles ro-vibrational calculations and observation in the laboratory, in the interstellar medium[118,119] and, most recently, in the atmosphere of Jupiter.[58,96] H_3^+ thus combines relative simplicity of its electronic motions (for example, it only has one truly bound electronic state) with great complexity of its nuclear motions.

The pioneering work on the ro-vibrational spectrum of H_3^+ and its isotopomers was done by Carney and Porter,[120–123] who calculated and fitted a potential energy surface which they then used for vibrational calculations of the low-lying levels. This work led directly to the first assigned spectra of H_3^+[124] and D_3^+[125] in 1980.

Since that time, work on these fundamental systems has intensified. There are a number of reasons for this. The lightness of the systems means that they are subject to extreme quantum behavior. Their floppiness makes both their vibrational and rotational motions difficult to represent by standard perturbation theory expansions based on the harmonic oscillator–rigid rotor model. Indeed, strong Coriolis effects has meant that spectra of the mixed isotopomers, H_2D^+ and D_2H^+, has defied assignment for a number of years.[126] This problem was only unraveled with the aid of sophisticated first principles calculations,[111,127–130] a process that is still continuing.[56,131,132]

Despite progress in assigning the infrared spectra there are still many unassigned lines in the observed discharge spectrum of all the isotopomers. Furthermore, even with available theoretical calculations,[50,98] little progress has been made in unraveling vibrational transitions involving high (>10) J states. Additionally, the pure rotational spectra of the mixed isotopomers are still poorly characterized. These spectra have additional significance because of the importance of H_3^+ and H_2D^+ in the chemistry of the interstellar medium.[133,134]

Besides the many spectra in the low-energy region, the really challenging problem from a theoretical point of view is the near dissociation spectra of Carrington and co-workers.[115,116] They have observed very dense spectra (over a 100 lines per cm^{-1}) involving quasi-bound states of all isotopomers of H_3^+. After many years, these spectra remain unassigned and largely unexplained. These experiments have spawned many classical studies and a number of quantum calculations, by ourselves[31,51,78,114] and other workers.[25,135–137] The complexity of this problem is such that none of the quantum-mechanical studies have even approached a full solution of this problem.

Our best efforts have involved calculating the highest truly bound rotational level for H_3^+ (about $J = 46$)[51] and using our DVR program to obtain the lowest 180 vibrational ($J = 0$) levels and $J = 1$ states for 40 of these.[31]

Table 2. Comparison of Calculated[(94)] and Observed[(141)] Dipole-Allowed Ro-Vibrational Transition Frequencies for H_3^+ in the ν_2 Fundamental[a] (with Calculated[(94)] Linestrengths and Einstein A Coefficients[b])

$J'G'_{U}$–$J''K''$	E_u (cm^{-1})	E_l (cm^{-1})	ω_{ul} (cm^{-1})	ω_{ul} (exp) (cm^{-1})[c]	$S(u-l)$ (Debye^2)	A_{ul} (s^{-1})
4 3_{-1} 5 3	3144.790	1080.139	2064.651	N.O.	0.330(−2)	0.101(+1)
4 2_{-1} 5 2	3259.663	1186.725	2072.938	N.O.	0.383(−2)	0.119(+1)
4 1_{-1} 5 1	3325.518	1249.910	2075.608	N.O.	0.132(−2)	0.410(+0)
4 5_{+1} 5 5	2863.711	728.777	2134.934	2134.922	0.238(+0)	0.808(+2)
4 4_{+1} 5 4	3069.003	928.672	2140.331	2140.348	0.190(+0)	0.648(+2)
4 3_{+1} 5 3	3233.027	1080.139	2152.888	2152.887	0.151(+3)	0.524(+2)
3 2_{-1} 4 2	2930.998	768.225	2162.773	N.O.	0.121(−2)	0.549(+0)
4 2_{+1} 5 2	3351.023	1186.725	2164.298	2164.278	0.126(+0)	0.444(+2)
3 1_{-1} 4 1	3002.486	833.311	2169.175	N.O.	0.776(−3)	0.356(+0)
4 1_{+1} 5 1	3422.758	1249.910	2172.847	2172.815	0.114(+0)	0.407(+2)
4 0_{-1} 5 0	3446.685	1270.862	2175.822	2175.780	0.110(+0)	0.397(+2)
3 4_{+1} 4 4	2719.308	501.875	2217.434	2217.451	0.184(+0)	0.897(+2)
3 3_{+1} 4 3	2876.591	658.503	2218.088	2218.129	0.137(+0)	0.670(+2)
3 2_{+1} 4 2	2992.161	768.225	2223.936	2223.965	0.104(+0)	0.513(+2)
3 1_{+1} 4 1	3063.193	833.311	2229.881	2229.895	0.857(−1)	0.426(+2)
2 1_{-1} 3 1	2755.291	494.606	2260.685	N.O.	0.551(−4)	0.400(−1)
2 1_{+1} 3 1	2790.127	494.606	2295.521	2295.577	0.607(−1)	0.460(+2)
2 2_{+1} 3 2	2723.765	427.881	2295.884	2295.947	0.864(−1)	0.656(+2)
2 0_{-1} 3 0	2812.642	516.713	2295.930	2295.980	0.523(−1)	0.396(+2)
2 3_{+1} 3 3	2614.135	315.248	2298.887	2298.930	0.130(+0)	0.992(+2)
1 1_{+1} 2 1	2609.377	237.279	2372.099	2372.185	0.386(−1)	0.540(+2)
1 2_{+1} 2 2	2548.041	169.246	2378.795	2378.869	0.775(−1)	0.109(+3)
0 1_{+1} 1 1	2521.282	64.104	2457.178	2457.290	0.257(−1)	0.119(+3)
5 4_{-1} 5 4	3395.911	928.672	2467.239	N.O.	0.865(−1)	0.370(+2)
5 0_{-1} 5 0	3742.346	1270.862	2471.483	2471.210	0.264(+0)	0.114(+3)
5 1_{-1} 5 1	3721.811	1249.910	2471.901	2472.325	0.254(+0)	0.109(+3)
5 3_{-1} 5 3	3552.620	1080.139	2472.481	2472.846	0.168(+0)	0.724(+2)
5 2_{-1} 5 2	3659.563	1186.725	2472.838	2473.238	0.222(+0)	0.955(+2)
4 3_{-1} 4 3	3144.790	658.503	2486.287	2486.559	0.855(−1)	0.458(+2)
4 2_{-1} 4 2	3259.663	768.225	2491.438	2491.749	0.162(+0)	0.871(+2)
4 1_{-1} 4 1	3325.518	833.311	2492.207	2492.541	0.206(+0)	0.111(+3)
3 2_{-1} 3 2	2930.998	427.881	2503.117	2503.347	0.850(−1)	0.597(+2)
3 1_{-1} 3 1	3002.486	494.606	2507.880	2508.131	0.152(+0)	0.107(+3)
3 0_{-1} 3 0	3025.528	516.713	2508.815	2509.075	0.174(+0)	0.123(+3)
2 1_{-1} 2 1	2755.291	237.279	2518.012	2518.207	0.837(−1)	0.838(+2)
1 0_{-1} 1 0	2616.487	86.933	2529.554	2529.724	0.761(−1)	0.129(+3)
1 1_{+1} 1 1	2609.377	64.104	2545.273	2545.418	0.385(−1)	0.663(+2)
2 1_{+1} 2 1	2790.127	237.279	2552.849	2552.987	0.216(−1)	0.226(+2)
2 2_{+1} 2 2	2723.765	169.246	2554.519	2554.664	0.430(−1)	0.250(+2)
3 3_{+1} 3 3	2876.591	315.248	2561.343	2561.493	0.454(−1)	0.341(+2)

Table 2. (*Continued*)

$J'G'_{U}$–$J''K''$	E_u (cm^{-1})	E_l (cm^{-1})	ω_{ul} (cm^{-1})	ω_{ul} (exp) (cm^{-1})	$S(u-l)$ (Debye^2)	A_{ul} (s^{-1})
3 2_{+1} 3 2	2992.161	427.881	2564.279	2564.408	0.324(−1)	0.244(+2)
4 4_{+1} 4 4	3069.003	501.875	2567.129	2567.285	0.470(−1)	0.278(+2)
3 1_{+1} 3 1	3063.193	494.606	2568.587	N.O.	0.851(−2)	0.774(+1)
5 5_{+1} 5 5	3299.725	728.777	2570.947	2571.111	0.481(−1)	0.234(+2)
4 3_{+1} 4 3	3233.027	658.503	2574.524	2574.660	0.385(−1)	0.229(+2)
5 4_{+1} 5 4	3509.703	928.672	2581.031	2581.184	0.410(−1)	0.201(+2)
4 2_{+1} 4 2	3531.023	768.225	2582.798	N.O.	0.162(−1)	0.974(+1)
4 1_{+1} 4 1	3422.758	833.311	2589.446	N.O.	0.378(−2)	0.229(+1)
5 3_{+1} 5 3	3673.497	1080.139	2593.358	2593.460	0.197(−1)	0.982(+1)
5 2_{+1} 5 2	3792.568	1186.725	2605.843	N.O.	0.845(−2)	0.426(+1)
5 1_{+1} 5 1	3862.944	1249.910	2613.034	N.O.	0.192(−2)	0.982(+0)
2 1_{-1} 1 1	2755.291	64.104	2691.187	2691.444	0.422(−1)	0.516(+2)
2 0_{-1} 1 0	2812.642	86.933	2725.709	2725.898	0.777(−1)	0.988(+2)
2 1_{+1} 1 1	2790.127	64.104	2726.023	2726.219	0.475(−1)	0.604(+2)
3 2_{-1} 2 2	2930.998	169.246	2761.752	2762.068	0.888(−1)	0.839(+2)
3 1_{-1} 2 1	3002.486	237.279	2765.207	2765.457	0.222(−1)	0.210(+2)
3 2_{+1} 2 2	2992.161	169.246	2822.915	2823.136	0.473(−1)	0.477(+2)
3 1_{+1} 2 1	3063.193	237.279	2825.914	2826.113	0.909(−1)	0.919(+2)
4 3_{-1} 3 3	3144.790	315.248	2829.542	2829.923	0.135(+0)	0.106(+3)
4 1_{-1} 3 1	3325.518	494.606	2830.912	2831.340	0.135(−1)	0.107(+1)
4 2_{-1} 3 2	3259.663	427.881	2831.782	2832.197	0.566(−1)	0.448(+2)
5 1_{-1} 4 1	3721.811	833.311	2888.500	N.O.	0.921(−2)	0.633(+1)
5 2_{-1} 4 2	3659.563	768.225	2891.338	2891.867	0.385(−1)	0.265(+2)
5 4_{-1} 4 4	3395.911	501.875	2894.037	2894.488	0.178(+0)	0.123(+3)
5 3_{-1} 4 3	3552.620	658.503	2894.117	2894.610	0.910(−1)	0.629(+2)
4 3_{+1} 3 3	3233.027	315.248	2917.779	2918.026	0.497(−1)	0.430(+2)
4 1_{+1} 3 1	3422.758	494.606	2928.152	2928.351	0.126(+0)	0.111(+3)
4 2_{+1} 3 2	3351.023	427.881	2923.142	2923.361	0.100(+0)	0.871(+2)
4 0_{-1} 3 0	3446.685	516.713	2929.972	2930.163	0.134(+0)	0.118(+3)
5 4_{+1} 4 4	3509.703	501.875	3007.828	3008.115	0.552(−1)	0.428(+2)
5 3_{+1} 4 3	3673.497	658.503	3014.994	3015.240	0.113(+0)	0.881(+2)
5 2_{+1} 4 2	3792.568	768.225	3024.343	3024.550	0.142(+0)	0.113(+3)
5 1_{+1} 4 1	3862.944	833.311	3029.633	3029.822	0.160(+0)	0.126(+3)

[a] Powers of ten in brackets.
[b] The Einstein A coefficients have been corrected for a factor $(2J_u + 1)$, mistakenly included in Ref. 94.
[c] N.O. = not experimentally observed.

Because the linear H_3^+ saddle point is already important for vibrational state 10, these calculations used spherical oscillator functions for the r_2 coordinate. Our calculations were the first three-dimensional quantum calculations to find "horseshoe" states of H_3^+; see Fig. 6. This classically stable orbit was discovered by Pollak and co-workers,[138–140] who used it to explain the observed coarse-grained structure of the Carrington–Kennedy spectrum. These spectra remain as a tantalizing reminder of what, even for superficially simple systems, remains to be done.

Unlike the quasi-bound levels of H_3^+, the low-lying levels have been very well characterized by first principles calculations. Table 2 gives a comparison between calculated and experimentally observed transition frequencies for H_3^+. Theoretical linestrengths and intensities are also given. Besides the impressive agreement between the frequencies, only two transitions with a calculated linestrength greater than 0.01 D^2 have not been observed. Conversely, no transitions weaker than this cut-off have been seen.

The success of first principles calculations on the fundamentals has led to concerted studies of the overtone, combination, and hot bands of H_3^+.[100,101] This work has led directly to the assignment of a number of H_3^+ hot bands[142] and a series of overtone transitions.[96] This latter collaboration between theory and experiment resulted in the assignment of a $2v_2(2)$ overtone emission spectrum in an auroral hot spot in Jupiter.[58] These observations had originally been undertaken to observe H_2 quadrupole lines but also saw a series of H_3^+ transitions. By our estimate the H_3^+ linestrengths are 10^9 larger than the H_2 transitions. Figure 7 compares the observed spectrum with our first principles spectrum. The observation that the intensities of the H_3^+ emission lines in Jupiter are time dependent[143,144] has spawned a host of observational studies of that planet.

One advantage of the variational calculation of ro-vibrational spectra is that it is possible to perform these calculations without taking any prior view of the outcome. Our intensity calculations contain no vibrational selection rules or rotational propensities. Furthermore, allowance is made for interaction, such as ro-vibrational coupling, which may relax the criteria usually used to identify strong transitions. We have thus been able to make detailed predictions for the "forbidden" rotational spectrum of H_3^+,[94] in broad agreement with previous phenomenological estimates,[145] and for a number of new, unsuspected bands that we believe to be observable with current techniques.[95]

7.2. The Alkali Cyanides

The alkali cyanides are very interesting prototypical floppy systems. Their unusual behavior in the condensed phase,[146] where the cyanide ions often behave as if they are spherical, has been studied for many years. Interest in the behavior of these systems in the gas phase was stimulated by a series of experiments on KCN,[147] NaCN,[148] and LiCN[149] performed in Nijmegen, the Netherlands. These studies showed that these molecules had

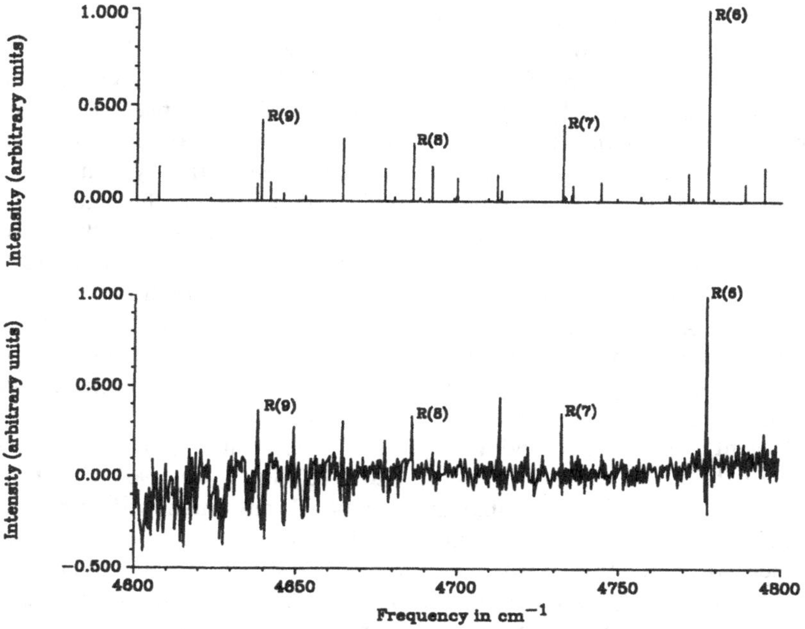

Figure 7. The simulated first principles H_3^+ $2v_2(2) \rightarrow v_0$ emission spectrum (top) compared with that measured by Drossart et al.[58] in the southern auroral "hot spot" on Jupiter (bottom). The emission line at 4712 cm^{-1} in the observed spectrum is due to an H_2 quadrupole transition.

unusual structures and properties and catalyzed parallel theoretical studies.

The two-dimensional (CN frozen) KCN surface of Wormer and Tennyson[150] showed a triangular equilibrium structure, in good agreement with experiment,[147] and a low barrier to linearity for the isocyanide KNC structure. A number of ro-vibrational studies were performed on this surface.[12–14,52,151,152] The upshot of these studies was that this system, despite being deeply bound, showed very wide amplitude motion which cannot be interpreted in conventional spectroscopic terms. Indeed, classical calculations[151,152] suggested that the strength of the mode coupling in KCN made the system almost pathologically chaotic. The onset of classical chaos is below the quantum ground state of the system, and in the chaotic region no classical periodic orbits could be found. Quantum-mechanical vibrational calculations show that the wave functions of the states do indeed reflect this behavior.[151,152]

The two-dimensional (CN frozen) LiCN surface of Essers et al.[153] shows the molecule to be linear. The surface has an absolute minimum for the isocyanide LiNC, in agreement with experiment,[149] and a metastable minimum for the cyanide LiCN structure. Again, the low barrier between these isomers has led to a number of ro-vibrational studies,[42,49,52,53,152,154] which are still continuing.[47,155] In contrast to KCN, this system has only weak mode coupling, leading to a later and slower onset of classical chaos.[152] The feeling that this is more typical of other isomerizing molecules[154] has stimulated the continuing interest in this system.

Properties of LiCN that have been studied include the fluorescence spectra of the lowest 80 states. These spectra and the related fluorescent lifetimes were found to have structures that could be closely related to the underlying classical dynamics of the system.[53] In particular, states that could be assigned (approximate) quantum numbers were found to behave very differently from those where no such assignment could be made. Subsequently, Brocks[156] found very similar behavior for RbCN, a molecule with a triangular equilibrium structure[157] like KCN.

Recently, Henderson and Tennyson[47] have been able to show the full power of the DVR method by using it to study the lowest 900 vibrational ($J = 0$) states of LiCN, diagonalizing a final Hamiltonian with dimension $L = 1870$. The majority of these states lie above the barrier to isomerization, and their wave functions show a rich variety of structures. Sample contour plots of some high-lying states are shown in Fig. 8. In the high-energy region, Henderson and Tennyson found normal-mode states localized about LiNC and LiCN, delocalized free-rotor states, and many irregular, delocalized states. The proportion of irregular states grew with energy but did not reach 100% for the energy region studied.

The molecule HCN shows some similarities with LiCN. Again, it has two linear minima but this time the stable isomer is HCN. The barrier in this system is higher, but more difficult for FBR calculations relying on polynomial functions. This is because H to CN center-of-mass distances are very different in the two isomers. HCN has been the subject of both two-dimensional[158] and three-dimensional[40] studies using our FBR procedures. However, the most comprehensive calculations on this system are undoubtedly the DVR calculations of Bačić and Light.[43]

7.3. Van der Waals Dimers

Van der Waals complexes form interesting examples of molecules that undergo large-amplitude motions. Indeed these systems are sometimes better understood in terms of models, such as the free rotor model, which start by assuming completely delocalized vibrational motion in the bending coordinate.[71,159]

Van der Waals complexes have played an important role in the development of methods for treating large-amplitude motions. Furthermore, the

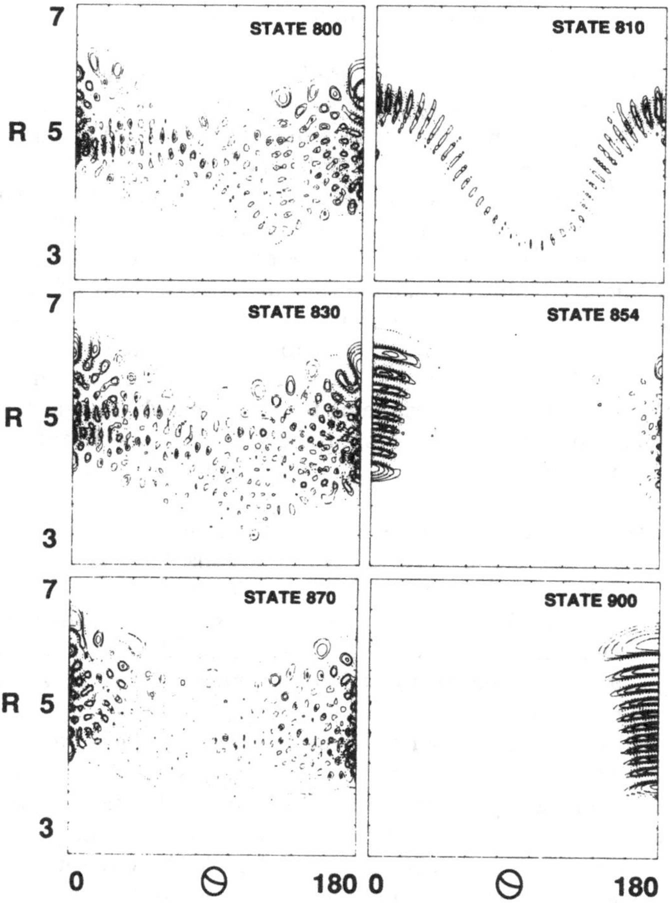

Figure 8. Contour plots of six high-lying LiCN states.[47] The plots are in scattering coordinates with the LiNC absolute minimum at 180°. The states on the left (numbers 800, 830, and 870) are unassigned; those on the right are free rotor (0, 73) (state 810), LiCN normal mode (17, 0), and LiNC normal mode (20, 0) (state 900). Solid (dashed) contours enclose regions where the wave function has positive (negative) amplitude. Contours are drawn at 4%, 8%, 16%, 32%, and 64% of the maximum amplitude of the wave function. The outer dashed contours represent the classical turning point of the potential for the associated eigenvalue.

relative simplicity of their potential surfaces has allowed the construction of an accurate potential energy surface by trial-and-error comparison of ro-vibrational calculation with observation. Well-characterized systems include the H_2–Noble gas complexes,[2] ArHCl,[160] and other Noble gas–hydrogen halide complexes.[161,162] Work on Van der Waals systems is still a very active area of investigation.[70,163]

Our methods rely heavily on methods developed to study Van der Waals complexes. Although this has not been the main focus of our research a number of calculations have been performed on Van der Waals systems such as H_2Ne,[14] H_3,[71] HeHF,[76] O_2Ar,[164] ArHCl,[165] ArCO,[87,166] O_2He,[167] and HeN_2^+.[159] These studies obtained estimates for all of the vibrational bound states of the system and some rotational excitation. The studies on HeHF and ArHCl explicitly included the effects of the diatom vibrational mode, and the O_2–Noble gas calculations included a study of fine structure and magnetic effects due to the O_2 spin.

The ArCO calculations mentioned above and some recent calculations on ArHCl by Choi and Light[46] considered the effect of highly excited rotational states. However, this seems to be a somewhat neglected area for Van der Waals systems. This is despite the fact that experiments performed at anything but the lowest temperatures may well access all the rotational states of a complex.

Recently García Allyón et al.[102] have performed calculations on the N_2Ar Van der Waals complex which explicitly considered *all* the truly bound states of the system. The calculations supplemented ones by Brocks[168] which only considered very low J levels and used an unrealistic model for the dipole function. García Allyón et al. calculated approximately 30,000 linestrengths and used the strongest 15,000 of these to synthesize the spectra shown in Figs. 3 and 4.

The temperature and Gaussian linewidths used to generate the spectra of Fig. 4 were chosen to mimic the conditions in the experiments of McKellar.[103] These experiments were the prime motivation for the calculations, and it is noteworthy that the calculations reproduce the observed experimental structure in the 20–45 cm^{-1} range.* These features are caused by the intensity from many Q branch transitions and cannot be properly understood without calculating a large number of rotational states. Features to higher frequency are undoubtedly caused by transitions involving quasi-bound states and thus were not reproduced by the model of García Allyón et al.

7.4. Metal Clusters

The study of small metal clusters is important in understanding the nucleation process in metals. They tend to have rather unusual structures,

* The experiments of McKellar were actually performed in the infrared and involved the simultaneous excitation of the N_2 vibrational mode. This was ignored in the analysis of García Allyón *et al.*, and comparison is thus made to experimental frequencies with the frequency of the N_2 vibration subtracted.

and their spectroscopy is so far rather poorly studied. Searles and co-workers[169–172] have performed a series of calculations on the Li_3^+ cluster, obtaining results in good agreement with ours.[173]

A more challenging problem is presented by the Na_3 cluster. This system shows some unusual experimental features,[174] which classical calculations suggest are due to the large-amplitude asymmetric stretching mode in the system.[175] The potential energy surface of Thompson *et al.* for Na_3, which is a Jahn–Teller system, shows a large number of critical points.[176] Besides the three symmetry-related absolute minima, there are three symmetry-related metastable minima and three linear saddle points. All these lie below the conical intersection which has an equilateral triangle geometry.

We[48] have performed a series of vibrational calculations on this system. Attempts to perform FBR calculations on this system gave poor results even for the low-lying levels of the system. By using Radau coordinates within a DVR framework we were able to converge results for the lowest 70 vibrational states of the system. An equivalent FBR calculation to this DVR calculation would have required in excess of 32,000 basis functions. The highest states studied are delocalized across the minima on the surface but lie below the linear saddle points and the conical intersection. Analysis of these results is still in progress.[177]

7.5. Hydrides

Another important and fundamental class of triatomic molecules that undergo large-amplitude vibration are those molecules containing hydrogen. The lightness of the hydrogen atom leads to enhanced quantum effects for hydrogen-containing systems.

The ro-vibrational levels of a whole range of hydrides have been the subject of accurate calculations over the years. Not surprisingly, the most popular of these systems has been water.[20] Variational calculations on water started in the very early days of ro-vibrational calculations.[3,4,178] The wealth of experimental data on this system has led to a number of potential energy surfaces[179,180] being constructed for this system by the trial-and-error matching of observed transitions to variational ro-vibrational calculations. Water is thus the only deeply bound polyatomic system besides H_3^+ for which the potential is known to spectroscopic accuracy. Studies of the related systems H_2S[57,181] and its isotopomers[182] have also been performed.

Calculations have also been performed on hydrogen-containing ions such as H_2F^+ and H_2Cl^+,[183] H_2He^+,[184] and CH_2^+.[20] As was mentioned in the introduction, CH_2^+ played an important part in the development of theories for treating large-amplitude motions. This system is quasi-linear, which makes it difficult to treat in a number of approaches. However, the motions are well represented by a bond length–bond angle coordinate system.[20] Because CH_2^+ is actually a Renner–Teller system most of these studies were performed to help the development of suitable theoretical methods for such systems rather than directly to aid spectroscopy. Only one calculation has actually tackled the full coupled-surface problem[19] and even this approximated the surface coupling terms.

The study of one rather unusual hydride—the molecule formed by the interaction of He with H_2 in its $B^1\Sigma_u^+$ excited electronic state—had some rather unexpected consequences. This system has two asymmetric, bent minima separated by a large barrier. HeH_2^* itself was well treated in scattering coordinates,[185] but scattering coordinate calculations on HeHD*, the system for which experimental information was available, failed. This was because the very different equilibrium geometries of the two isotopomers, HeHD* and HeDH*, when expressed from the HD* center-of-mass, made it impossible to develop radial basis functions that adequately covered both potential minima.

This led to a search for other coordinate systems. The result was Sutcliffe and Tennyson's generalized coordinates.[32,37] The particular coordinates suited to calculations on HeHD* were christened geometric coordinates ($g_1 = 0.5$, $g_2 = 0.0$); see Table 1. These nonorthogonal coordinates use the geometric center of the HD* bond rather than the center-of-mass. They are thus geometrically the same as the scattering coordinates used for the HeH_2^* problem. In these coordinates the symmetry mixing introduced into an AB_2-type problem upon isotopic substitution of one of the A atoms is carried by off-diagonal terms in the kinetic energy operator. We would argue that this is much more physical than introducing coordinates which cause the potential, which is still symmetric within the Born–Oppenheimer approximation, to appear asymmetric.

8. Conclusions

In this chapter we have discussed the progress that has been made in calculating ro-vibrational states of triatomic systems during the last decade. Given a potential energy surface, the fundamental vibrational levels and low-lying rotational states of any triatomic system can now be accurately determined using variational techniques. Attention for triatomic systems has thus turned to the development of techniques for studying high-lying ro-vibrational states. For Van der Waals systems all the bound vibrational states for a given potential energy surface can already be accurately determined.

The two-step variational techniques of Chen et al.[22] and Tennyson and Sutcliffe[50] would appear to have effectively solved the rotational excitation problem. It is to be hoped that theoretical advances, and in particular the use of discrete variable representations (DVRs),[45] will similarly lead to methods that can determine all the vibrational bound states of any given triatomic problem. The few systems for which high-lying levels have already been determined have indicated a great richness in the solutions found and hinted at unusual features waiting to be observed.

Finally, in this chapter we have confined our discussion to triatomic molecules. This does not indicate that no worthwhile calculations have been performed on larger systems. A number of full variational calculations are available for tetratomic molecules; however, these calculations are still difficult to perform. It is probably fair to say that no clear pattern has yet emerged as to what will prove to be the best, or most enduring, method for doing these calculations.[186]

Acknowledgments

We gladly acknowledge the contribution of Brian Sutcliffe with whom we have collaborated for many years and much of whose work is reproduced here. Parts of this work have been supported by the Science and Engineering Research Council under grants GR/D/93407, GR/E/45021, GR/E/98348, and GR/F/14550, and by the Research Corporation Trust.

References

1. G. D. Carney, L. L. Sprandel, and C. W. Kern, *Advan. Chem. Phys.* **37**, 305–379 (1978).
2. R. J. Le Roy and J. S. Carley, *Advan. Chem. Phys.* **42**, 353–420 (1980).
3. R. J. Whitehead and N. C. Handy, *J. Mol. Spectrosc.* **55**, 356–373 (1975).
4. R. J. Whitehead and N. C. Handy, *J. Mol. Spectrosc.* **59**, 459–469 (1976).
5. C. Eckart, *Phys. Rev.* **47**, 552–558 (1935).
6. J. K. G. Watson, *Mol. Phys.* **15**, 479–490 (1968).
7. E. C. K. Lai, Master's Thesis, Department of Chemistry, Indiana University (1975).
8. R. J. Le Roy and J. Van Kranendonk, *J. Chem. Phys.* **61**, 4750–4769 (1974).
9. P. Jensen, *Computer Phys. Reports* **1**, 1–56 (1983).
10. J. Tennyson, *Computer Phys. Reports* **4**, 1–36 (1986).
11. S. Carter and N. C. Handy, *Computer Phys. Reports* **5**, 115–172 (1986).
12. J. Tennyson and B. T. Sutcliffe, *Mol. Phys.* **46**, 97–109 (1982).
13. J. Tennyson and A. van der Avoird, *J. Chem. Phys.* **76**, 5710–5718 (1982).
14. J. Tennyson and B. T. Sutcliffe, *J. Chem. Phys.* **77**, 4061–4072 (1982).
15. R. Bartholomae, D. Martin, and B. T. Sutcliffe, *J. Mol. Spectrosc.* **87**, 367–381 (1981).
16. S. Carter and N. C. Handy, *J. Mol. Spectrosc.* **95**, 9–19 (1982).

17. J. Tennyson and B. T. Sutcliffe, *J. Mol. Spectrosc.* **101**, 71–82 (1983).
18. S. Carter, N. C. Handy, and B. T. Sutcliffe, *Mol. Phys.* **49**, 745–748 (1983).
19. S. Carter and N. C. Handy, *Mol. Phys.* **52**, 1367–1391 (1984).
20. B. T. Sutcliffe and J. Tennyson, *J. Chem. Soc., Faraday Trans. 2* **83**, 1663–1674 (1987).
21. J. S. Lee and D. Secrest, *J. Chem. Phys.* **92**, 1821–1830 (1989).
22. C.-L. Chen, B. Maessen, and M. Wolfsberg, *J. Chem. Phys.* **83**, 1795–1807 (1985).
23. G. Brocks, A. van der Avoird, B. T. Sutcliffe, and J. Tennyson, *Mol. Phys.* **50**, 1025–1043 (1983).
24. P. Jensen, *J. Mol. Spectrosc.* **128**, 478–501 (1988).
25. R. M. Whitnell and J. C. Light, *J. Chem. Phys.* **90**, 1774–1786 (1989).
26. B. T. Sutcliffe, in: *Current Aspects of Quantum Chemistry* (R. Carbo, ed.), Studies in Theoretical Chemistry, Vol. 21, pp. 99–125, Elsevier, New York (1982).
27. N. C. Handy, *Mol. Phys.* **61**, 207–223 (1987).
28. J. Tennyson, *Computer Phys. Comms.* **29**, 307–319 (1983).
29. J. Tennyson, *Computer Phys. Comms.* **32**, 109–114 (1984).
30. J. Tennyson and B. T. Sutcliffe, *Mol. Phys.* **51**, 887–906 (1984).
31. J. Tennyson and J. R. Henderson, *J. Chem. Phys.* **91**, 3815–3825 (1989).
32. B. T. Sutcliffe and J. Tennyson, *Mol. Phys.* **58**, 1053–1066 (1986).
33. D. Estes and D. Secrest, *Mol. Phys.* **59**, 569–578 (1986).
34. J. Makarewicz, *J. Phys. B: At. Mol. Opt. Phys.* **21**, 1803–1819 (1988).
35. J. Makarewicz and W. Lodyga, *Mol. Phys.* **64**, 899–919 (1988).
36. J. M. Bowman, J. Zuniga, and A. Wierzbicki, *J. Chem. Phys.* **90**, 2708–2713 (1989).
37. B. T. Sutcliffe and J. Tennyson, *Int. J. Quantum Chem.* **29**, 183–196 (1991).
38. J. Tennyson, *Computer Phys. Comms.* **42**, 257–270 (1986).
39. J. Tennyson and S. Miller. *Computer Phys. Comms.* **55**, 149–175 (1989).
40. B. T. Sutcliffe and J. Tennyson, in: *Molecules in Physics, Chemistry, and Biology* (J. Maruani, ed.), Vol. II, pp. 313–333, Kluwer, Dordrecht (1988).
41. J. C. Light, I. P. Hamilton, and J. V. Lill, *J. Chem. Phys.* **92**, 1400–1409 (1985).
42. Z. Bačić and J. C. Light, *J. Chem. Phys.* **85**, 4594–4604 (1986).
43. Z. Bačić and J. C. Light, *J. Chem. Phys.* **86**, 3065–3077 (1987).
44. Z. Bačić, D. Watt, and J. C. Light, *J. Chem. Phys.* **89**, 947–955 (1988).
45. Z. Bačić and J. C. Light, *Ann. Rev. Phys. Chem.* **40**, 469–498 (1989).
46. S. E. Choi and J. C. Light, *J. Chem. Phys.* **92**, 2129–2145 (1990).
47. J. R. Henderson and J. Tennyson, *Mol. Phys.* **69**, 639–648 (1990).
48. J. R. Henderson, S. Miller, and J. Tennyson, *J. Chem. Soc., Faraday Trans.* **86**, 1963–1968 (1990).
49. G. Brocks and J. Tennyson, *J. Mol. Spectrosc.* **99**, 263–278 (1983).
50. J. Tennyson and B. T. Sutcliffe, *Mol. Phys.* **58**, 1067–1085 (1986).
51. S. Miller and J. Tennyson, *Chem. Phys. Letts.* **145**, 117–120 (1988).
52. G. Brocks, J. Tennyson, and A. van der Avoird, *J. Chem. Phys.* **80**, 3223–3233 (1984).
53. J. Tennyson, G. Brocks, and S. C. Farantos, *Chem. Phys.* **104**, 399–407 (1986).
54. P. Jensen, *J. Mol. Spectrosc.* **132**, 429–457 (1988).
55. S. Carter, J. Senekowitsch, N. C. Handy, and P. Rosmus, *Mol. Phys.* **65**, 143–160 (1988).
56. S. Miller, J. Tennyson, and B. T. Sutcliffe, *Mol. Phys.* **66**, 429–456 (1989).
57. S. Carter, P. Rosmus, N. C. Handy, S. Miller, J. Tennyson, and B. T. Sutcliffe, *Computer Phys. Comms.* **55**, 71–75 (1989).
58. P. Drossart, J.-P. Maillard, J. Caldwell, S. J. Kim, J. K. G. Watson, W. A. Majewski, J. Tennyson, S. Miller, S. Atreya, J. Clarke, J. H. Waite, Jr., and R. Wagener, *Nature (London)* **340**, 539–541 (1989).
59. B. R. Johnson and W. P. Reinhardt, *J. Chem. Phys.* **85**, 4538–4556 (1986).

60. D. M. Brink and G. R. Satchler, *Angular Momentum*, 2nd ed., Clarendon, Oxford (1968).
61. A. M. Arthurs and A. Dalgarno, *Proc. Roy. Soc. Ser. A* **256**, 540–551 (1960).
62. E. U. Condon and G. H. Shortley, *The Theory of Atomic Spectra*, Cambridge University Press, Cambridge (1935).
63. B. T. Sutcliffe, *Mol. Phys.* **48**, 561–566 (1983).
64. A. H. Stroud and D. Secrest, *Gaussian Quadrature Formulas*, Prentice-Hall, London (1966).
65. J. Tennyson, *Computer Phys. Comms.* **36**, 39–41 (1985).
66. J. M. Hutson, in: *Advances in Molecular Vibrations and Collision Dynamics, Vol. 1* (J. M. Bowman, ed.), p. 1 (1991).
67. J. Tennyson, *Chem. Phys. Letts.* **86**, 181–184 (1982).
68. D. Cropek and G. D. Carney, *J. Chem. Phys.* **80**, 4280–4285 (1984).
69. D. J. Searles and E. I. von Nagy-Felsobuki, *Am. J. Phys.* **56**, 444–448 (1988).
70. P. M. Morse, *Phys. Rev.* **34**, 57–65 (1935).
71. D. ter Haar, *Phys. Rev.* **70**, 222–223 (1946).
72. J. Tennyson and B. T. Sutcliffe, *J. Chem. Phys.* **79**, 43–51 (1983).
73. I. S. Gradshteyn and I. H. Ryzhik, *Tables of Integrals, Series, and Products*, Academic, New York (1980).
74. J. Tennyson, O. Brass, and E. Pollak, *J. Chem. Phys.* **92**, 3005–3017 (1990).
75. D. O. Harris, G. O. Engerholm, and W. Gwinn, *J. Chem. Phys.* **43**, 1515–1517 (1965).
76. A. S. Dickinson and P. R. Certain, *J. Chem. Phys.* **49**, 4209–4211 (1968).
77. A. C. Peat and W. Yang, *J. Chem. Phys.* **90**, 1746–1751 (1989).
78. S. Carter and N. C. Handy, *Mol. Phys.* **57**, 175–185 (1986).
79. J. C. Light, R. M. Whitnell, T. J. Pack, and S. E. Choi, in: *Supercomputer Algorithms for Reactivity, Dynamics, and Kinetics of Small Molecules* (A. Laganà, ed.), NATO ASI Series C, Vol. 277, pp. 187–213, Kluwer, Dordrecht (1989).
80. S. Miller and J. Tennyson, unpublished work.
81. P. R. Bunker, *Molecular Symmetry and Spectroscopy*, Academic, New York (1979).
82. G. S. Ezra, *Symmetry Properties of Molecules*, Lecture Notes in Chemistry, Vol. 28, Springer-Verlag, Berlin (1982).
83. J. M. Brown, J. T. Hougen, K.-P. Huber, J. W. C. Johns, I. Kopp, H. Lefrebvre-Brion, A. J. Merer, D. A. Ramsay, J. Rostos, and R. N. Zara, *J. Mol. Spectrosc.* **55**, 500–501 (1975).
84. J. Tennyson, S. Miller, and B. T. Sutcliffe, in: *Supercomputer Algorithms for Reactivity, Dynamics, and Kinetics of Small Molecules* (A. Laganà, ed.), NATO ASI Series C, Vol. 277, pp. 261–270, Kluwer, Dordrecht (1989).
85. R. M. Whitnell and J. C. Light, *J. Chem. Phys.* **89**, 3674–3680 (1988).
86. S. L. Holmgren, M. Waldman, and W. Klemperer, *J. Chem. Phys.* **67**, 4414–4422 (1977).
87. J. Tennyson, S. Miller, and B. T. Sutcliffe, *J. Chem. Soc., Faraday Trans. 2* **84**, 1295–1303 (1988).
88. B. T. Sutcliffe, S. Miller, and J. Tennyson, *Computer Phys. Comms.* **51**, 73–82 (1988).
89. B. T. Sutcliffe, J. Tennyson, and S. Miller, *Theor. Chim. Acta* **72**, 265–276 (1987).
90. V. Špirko, P. Jensen, P. R. Bunker, and A. Čejchan, *J. Mol. Spectrosc.* **112**, 183–212 (1985).
91. P. Jensen, V. Špirko, and P. R. Bunker, *J. Mol. Spectrosc.* **115**, 269–293 (1986).
92. S. Carter and N. C. Handy, *J. Chem. Phys.* **87**, 4294–4301 (1987).
93. W. L. Meerts, private communication (1987).
94. S. Miller and J. Tennyson, *Astrophys. J.* **335**, 486–490 (1988).
95. S. Miller, J. Tennyson, and B. T. Sutcliffe, *J. Mol. Spectrosc.* **141**, 104–117 (1990).
96. W. A. Majewski, P. A. Feldman, J K. G. Watson, S. Miller, and J. Tennyson, *Astrophys. J.* **347**, L51–L54 (1989).

97. B. Weis, S. Carter, P. Rosmus, H.-J. Werner, and P. J. Knowles, *J. Chem. Phys.* **91**, 2818–2833 (1989)
98. P. Jensen and V. Špirko, *J. Mol. Spectrosc.* **118**, 208–231 (1986).
99. S. Miller and J. Tennyson, *J. Mol. Spectrosc.* **126**, 183–192 (1987).
100. S. Miller and J. Tennyson, *J. Mol. Spectrosc.* **128**, 530–539 (1988); **133**, 237 (1989).
101. S. Miller and J. Tennyson, *J. Mol. Spectrosc.* **136**, 223–240 (1989).
102. A. García Allyón, J. Santamaria, S. Miller, and J. Tennyson, *Mol. Phys.* **71**, 1043–1054 (1990).
103. A. R. W. McKellar, *J. Chem. Phys.* **88**, 4190–4196 (1988).
104. J. M. Hutson and R. J. Le Roy, *J. Chem. Phys.* **83**, 1197–1203 (1985).
105. J. M. Hutson, *J. Chem. Soc., Faraday Trans. 2* **82**, 1163–1171 (1986).
106. J. M. Bowman, J. S. Bitman, and L. B. Harding, *J. Chem. Phys.* **85**, 911–921 (1986).
107. B. S. Garbow, J. M. Boyle, J. J. Dongarra, and C. B. Moler, *Matrix Eigensystem Routines: EISPACK Guide Extension*, Lecture Notes in Computer Science, Vol. 51, Springer-Verlag, New York (1977).
108. J. Tennyson and A. van der Avoird, *J. Chem. Phys.* **77**, 5664–5681 (1982); **80**, 2986 (1984).
109. *NAG Fortran Library Manual*, Mark 11, Vol. 4 (1983).
110. P. J. Nikolai, *ACM Trans. Math. Software* **5**, 118–128 (1979).
111. J. Tennyson and B. T. Sutcliffe, *Mol. Phys.* **54**, 141–144 (1985).
112. J. Tennyson and B. T. Sutcliffe, *Mol. Phys.* **56**, 1175–1183 (1985).
113. J. Tennyson and B. T. Sutcliffe, *J. Chem. Soc., Faraday Trans. 2* **82**, 1151–1162 (1986).
114. O. Brass, J. Tennyson, and E. Pollak, *J. Chem. Phys.* **92**, 3377–3386 (1990).
115. A. Carrington, J. Buttenshaw, and R. A. Kennedy, *Mol. Phys.* **45**, 753–758 (1982).
116. A. Carrington and R. A. Kennedy, *J. Chem. Phys.* **81**, 91–112 (1984).
117. W. Meyer, P. Botschwina, and P. G. Burton, *J. Chem. Phys.* **84**, 891–900 (1986).
118. T. Oka, *Phil. Trans. R. Soc. Lond. A* **303**, 543–549 (1981).
119. T. R. Geballe and T. Oka, *Astrophys. J.* **342**, 855–859 (1989).
120. G. D. Carney and R. N. Porter, *J. Chem. Phys.* **60**, 4251–4264 (1974).
121. G. D. Carney and R. N. Porter, *J. Chem. Phys.* **65**, 3547–3565 (1976).
122. G. D. Carney and R. N. Porter, *Chem. Phys. Letts.* **50**, 327–329 (1977).
123. G. D. Carney and R. N. Porter, *Phys. Rev. Letts.* **45**, 537–541 (1980).
124. T. Oka, *Phys. Rev. Letts.* **45**, 531–534 (1980).
125. J.-T. Shy, J. W. Farley, W. E. Lamb, Jr., and W. H. Wing, *Phys. Rev. Letts.* **45**, 535–537 (1980).
126. J.-T. Shy, J. W. Farley, and W. H. Wing, *Phys. Rev. A* **24**, 1146–1149 (1981).
127. G. D. Carney, *Chem. Phys.* **54**, 103–107 (1980).
128. G. D. Carney, *Can. J. Phys.* **62**, 1871–1874 (1984).
129. S. C. Foster, A. R. W. McKellar, I. R. Peterkin, J. K. G. Watson, F. S. Pan, M. W. Crofton, R. S. Altman, and T. Oka, *J. Chem. Phys.* **84**, 91–99 (1986).
130. S. C. Foster, A. R. W. McKellar, and J. K. G. Watson, *J. Chem. Phys.* **85**, 664–670 (1986).
131. I. N. Kozin, O. L. Polyansky, and N. F. Zobov, *J. Mol. Spectrosc.* **128**, 126–134 (1988).
132. O. L. Polyansky and A. R. W. McKellar, *J. Chem. Phys.* **92**, 4039–4043.
133. E. Herbst and W. Klemperer, *Astrophys. J.* **185**, 505–533 (1973).
134. E. F. van Dishoeck, in: *Proc. Conf. Space-Borne Sub-Millimetre Astronomy Mission*, Segovia, Spain, p. 107 (1986).
135. R. Pfeiffer and M. S. Child, *Mol. Phys.* **60**, 1367–1378 (1987).
136. J. M. Gomez Llorrente, J. Zakrzewski, H. S. Taylor, and K. C. Kulander, *J. Chem. Phys.* **89**, 5959–5960 (1989).

137. J. M. Gomez Llorrente, J. Zakrzewski, H. S. Taylor, and K. C. Kulander, *J. Chem. Phys.* **90**, 1505–1518 (1989).
138. M. Berlinger, E. Pollak, and C. Schlier, *J. Chem. Phys.* **88**, 5643–5656 (1988).
139. J. M. Gomez Llorrente and E. Pollak, *J. Chem. Phys.* **89**, 1195–1196 (1989).
140. J. M. Gomez Llorrente and E. Pollak, *J. Chem. Phys.* **90**, 5406–5419 (1989).
141. W. A. Majewski, M. D. Marshall, A. R. W. McKellar, J. W. C. Johns, and J. K. G. Watson, *J. Molec. Spectrosc.* **122**, 567–582 (1987).
142. M. G. Bawendi, B. D. Rehfuss, and T. Oka, *J. Chem. Phys.* **93**, 6200–6209 (1990).
143. L. Trafton, D. F. Lester, and K. L. Thomson, *Astrophys. J.* **343**, L73–L76 (1989).
144. T. Oka and T. R. Geballe, *Astrophys J.* **351**, L53–L57 (1990).
145. F.-S. Pan and T. Oka, *Astrophys. J.* **305**, 518–525 (1986).
146. S. Miller and J. H. R. Clarke, *J. Chem. Soc., Faraday Trans. 2* **74**, 160–173 (1978).
147. J. J. van Vaals, W. L. Meerts, and A. Dymanus, *J. Mol. Spectrosc.* **106**, 280–298 (1984).
148. J. J. van Vaals, W. L. Meerts, and A. Dymanus, *Chem. Phys.* **80**, 147–159 (1984).
149. J. J. van Vaals, W. L. Meerts, and A. Dymanus, *Chem. Phys.* **82**, 385–393 (1983).
150. P. E. S. Wormer and J. Tennyson, *J. Chem. Phys.* **75**, 1245–1252 (1981).
151. J. Tennyson and S. C. Farantos, *Chem. Phys. Letts.* **109**, 160–165 (1984).
152. S. C. Farantos and J. Tennyson, *J. Chem. Phys.* **82**, 800–809 (1985).
153. R. Essers, J. Tennyson, and P. E. S. Wormer, *Chem. Phys. Letts.* **89**, 223–227 (1982).
154. J. Tennyson and S. C. Farantos, *Chem. Phys.* **93**, 237–244 (1985).
155. R. M. Benito, F. Borondo, J.-H. Kim, B. G. Sumpter, and G. S. Ezra, *Chem. Phys. Letts.* **161**, 60–65 (1989).
156. G. Brocks, *Chem. Phys.* **116**, 33–34 (1987).
157. E. van Leuken, G. Brocks, and P. E. S. Wormer, *Chem. Phys.* **110**, 365–373 (1986).
158. M. Founargiotakis, S. C. Farantos, and J. Tennyson, *J. Chem. Phys.* **88**, 1598–1607 (1988).
159. S. Miller, J. Tennyson, B. Follmeg, P. Rosmus, and H.-J. Werner, *J. Chem. Phys.* **89**, 2178–2184 (1988).
160. J. M. Hutson, *J. Chem. Phys.* **89**, 4550–4557 (1988).
161. J. M. Hutson, *J. Chem. Phys.* **91**, 4448–4454 (1989).
162. J. M. Hutson, *J. Chem. Phys.* **91**, 4455–4461 (1989).
163. *Dynamics of Polyatomic Van der Waals Complexes* (N. Halberstadt and K. Janda, eds.), NATO ASI Series B, Plenum, New York (1990).
164. J. Tennyson and J. Mettes, *Chem. Phys.* **76**, 195–202 (1983).
165. J. Tennyson, *Mol. Phys.* **55**, 463–473 (1985).
166. S. C. Farantos and J. Tennyson, in: *New Concepts in Physical Chemistry* (L. S. Cederbaum, A. Mann, and W. Gans, eds.), pp. 195–206, Reidel, Dordrecht (1988).
167. J. Tennyson and A. van der Avoird, *Chem. Phys. Letts.* **105**, 49–53 (1984).
168. G. Brocks, Ph.D. Thesis, University of Nijmegen (1987).
169. S. J. Dunne, D. J. Searles, and E. I. von Nagy-Felsobuki, *Spectrochim. Acta* **43A**, 699–701 (1987).
170. D. J. Searles, S. J. Dunne, and E. I. von Nagy-Felsobuki, *Spectrochim. Acta* **44A**, 505–515 (1988).
171. D. J. Searles, S. J. Dunne, and E. I. von Nagy-Felsobuki, *Spectrochim. Acta* **44A**, 985–989 (1988).
172. D. J. Searles and E. I. von Nagy-Felsobuki, *Aust. J. Chem.* **42**, 737–739 (1989).
173. J. R. Henderson, S. Miller, and J. Tennyson, *Spectrochim. Acta* **44A**, 1287–1290 (1988).
174. M. Broyer, G. Delacrétaz, G.-Q. Ni, R. L. Whettan, J. P. Wolfe, and L. Wöste, *J. Chem. Phys.* **90**, 4620–4622 (1989).
175. J. M. Gomez Llorrente and H. S. Taylor, *J. Chem. Phys.* **91**, 953–962 (1989).

176. T. C. Thompson, G. Izmirlian, Jr., S. J. Lemon, D. G. Truhlar, and C. A. Mead, *J. Chem. Phys.* **82**, 5597–5603 (1985).
177. N. G. Fulton, S. Miller, J. R. Henderson, and J. Tennyson (unpublished).
178. M. G. Bucknell, N. C. Handy, and S. F. Boys, *Mol. Phys.* **28**, 759–776 (1974).
179. S. Carter and N. C. Handy, *J. Chem. Phys.* **87**, 4294–4301 (1987).
180. P. Jensen, *J. Mol. Spectrosc.* **133**, 438–460 (1989).
181. J. Senekowitsch, S. Carter, A. Zilch, H.-J. Werner, N. C. Handy, and P. Rosmus, *J. Chem. Phys.* **90**, 783–794 (1989).
182. S. Miller, J. Tennyson, P. Rosmus, J. Senekowitsch, and I. M. Mills, *J. Mol. Spectrosc.* **143**, 61–80 (1990).
183. P. Wells and J. Tennyson, unpublished work.
184. J. Tennyson and S. Miller, *J. Chem. Phys.* **87**, 6648–6652 (1987).
185. S. C. Farantos and J. Tennyson, *J. Chem. Phys.* **82**, 2163 (1985).
186. J. Tennyson, in: *Dynamics of Polyatomic Van der Waals Complexes* (N. Halberstadt and K. Janda, eds.), NATO ASI Series B, Plenum, New York (1990).
187. Tennyson and B. T. Sutcliffe, *Int. J. Quantum Chem.* (in press).

4

The Many-Body Perturbation Theory of the Vibrational–Electronic Problem in Molecules

IVAN HUBAČ AND MICHAL SVRČEK

1. Introduction

The analytical evaluation of energy derivatives is one of the foremost achievements of modern ab initio quantum chemistry.[1-3] Derivatives of the energy with respect to external field perturbation parameters are directly associated with the one-electron properties of the molecule such as moments, field gradients, spin densities, etc. (for details, see Ref. 1).

Consider the Schrödinger equation corresponding to a molecular system, where the electronic Hamiltonian $H(\chi)$, its eigenfunctions $\psi_i(\chi)$, and its eigenvalues $E_i(\chi)$ are dependent upon some parameter χ[2]

$$H(\chi)\psi_i(\chi) = E_i(\chi)\psi_i(\chi) \tag{1.1}$$

$$H(\chi) = -\tfrac{1}{2}\sum_i \nabla_i^2 - \sum_{i,a} Z_a r_{ai}^{-1} + \sum_i \chi w_i + \sum_{i>j} r_{ij}^{-1} \tag{1.2}$$

IVAN HUBAČ • Division of Chemical Physics, Faculty of Mathematics and Physics, Komensky University, 842 15 Bratislava, Czechoslovakia. MICHAL SVRČEK • Institute of Chemistry, Division of Chemistry and Physics of Biomolecules, Komensky University, 832 33 Bratislava, Czechoslovakia.

Methods in Computational Chemistry, Volume 4: Molecular Vibrations, edited by Stephen Wilson. Plenum Press, New York, 1992.

or

$$H(\chi) = h + v + \sum_i \chi w_i \tag{1.3}$$

where h is the one-particle operator

$$h = -\tfrac{1}{2}\sum_i \nabla_i^2 - \sum_{i,a} Z_a r_{ai}^{-1} \tag{1.4}$$

and v is the two-particle operator

$$v = \sum_{i>j} r_{ij}^{-1} \tag{1.5}$$

The parameter χ may control the displacement of an atomic nucleus in the molecular system, or it may be the strength of an external one-electron potential energy perturbation w_i, such as a uniform electric field applied in the z direction ($\chi w_i = \chi Z_i$).

Using the formalism of second quantization[4] we can write the Hamiltonian (1.3) in the form

$$H(\chi) = \sum_{PQ} \langle P|h|Q\rangle a_P^+ a_Q + \tfrac{1}{2} \sum_{PQRS} \langle PQ|v|RS\rangle a_P^+ a_Q^+ a_S a_R + \sum_{PQ} \left\langle P \middle| \sum_i \chi w_i \middle| Q \right\rangle a_P^+ a_Q \tag{1.6}$$

where $a_P^+(a_Q)$ are the creation (annihilation) operators defined on the one-particle basis set $\{|P\rangle, |Q\rangle, \ldots\}$. Using the concept of normal product (N-product) we can rewrite the Hamiltonian (1.6) into the N-product form[2,4]

$$\begin{aligned} H_N(\chi) = {} & \langle \Phi_0(\chi)|H(\chi)|\Phi_0(\chi)\rangle + \sum_{PQ} \langle P(\chi)|f(\chi)|Q(\chi)\rangle N[a_P^+(\chi)a_Q(\chi)] \\ & + \tfrac{1}{2} \sum_{PQRS} \langle P(\chi)Q(\chi)|v(\chi)|R(\chi)S(\chi)\rangle \\ & \cdot N[a_P^+(\chi)a_Q^+(\chi)a_S(\chi)a_R(\chi)] \end{aligned} \tag{1.7}$$

where $N[\ ]$ is the normal product of creation ($a_P^+(\chi)$) and annihilation $a_Q(\chi)$) operators, $|\Phi_0(\chi)\rangle$ is the χ-dependent reference state vector, and f is the χ-dependent Hartree–Fock operator. In the last two decades, the many-body methods, for example, many-body perturbation theory (MBPT) as well as coupled cluster (CC) methods,[5-13] have become very successful in the calculation of the correlation energies of atomic or molecular systems: That is, they have permitted calculations beyond the Hartree–Fock method with very high efficiency. Therefore, it has become important to introduce the analytic gradient techniques into MBPT and/or the CC methods (see, e.g., Ref. 2 and references therein). The important step in this respect was the introduction of an efficient method for the solution of the coupled perturbed Hartree–Fock (CPHF) equations.[14-16] Pople *et al.*[16] deal with the calculation of MP2 (second order of MBPT) derivatives. We are not going to discuss here the algorithm for how to calculate the MP2 gradients. The efficient way to calculate these quantities is discussed in other papers (see, e.g., Refs. 1 and 17). Evaluation of gradients in the CC method and in higher orders of MBPT is discussed in Refs. 2 and 18–19. Recently, the procedure for the calculation of Moller–Plesset energy derivatives based on the theory of Lagrangians was developed by Jorgensen.[20,21]. In our contribution we present an alternative approach based on the technique of canonical transformations. To explain our approach more clearly let us imagine a purely electronic Hamiltonian without the presence of the external perturbation $\sum_i \chi w_i$:

$$H^e = z + v \tag{1.8}$$

The N-product form of this Hamiltonian is

$$H_N^e = \langle\Phi_0|H^e|\Phi_0\rangle + \sum_{PQ} \langle P|f|Q\rangle N[a_P^+ a_Q] + \frac{1}{2} \sum_{PQRS} \langle PQ|v|RS\rangle N[a_P^+ a_Q^+ a_S a_R] \tag{1.9}$$

where f is the Hartree–Fock operator.

If we ask our basis set $\{|P\rangle\}$ to be the solution of the Hartree–Fock equations

$$f|P\rangle = \varepsilon_P|P\rangle \tag{1.10}$$

and split the Hamiltonian (1.9) into an unperturbed part H_0 and the perturbation H'

$$H_N^e = \epsilon_0 + H_0 + H' \tag{1.11}$$

where

$$\epsilon_0 = \langle\Phi_0|H^e|\Phi_0\rangle \tag{1.12}$$

$$H_0 = \sum_P \varepsilon_P N[a_P^+ a_P] \tag{1.13}$$

and

$$H' = \frac{1}{2} \sum_{PQRS} \langle PQ|v|RS\rangle N[a_P^+ a_Q^+ a_S a_R] \tag{1.14}$$

then we have very well known Moller–Plesset splitting of the Hamiltonian shown in (1.9).[22] In the last two decades this Hamiltonian has been very successfully used for the calculation of the correlation energies of atoms and molecules either using MBPT or CC methods.[5–13] In MBPT we proceed with the perturbed (exact) Schrödinger equation

$$H_N^e \psi_i = E_i \psi_i \tag{1.15}$$

and the unperturbed Schrödinger equation

$$H_0|\Phi_i\rangle = e_i|\Phi_i\rangle \tag{1.16}$$

The eigenvalue E_i can be found using the nondegenerate Rayleigh–Schrödinger perturbation expansion in the form

$$E_i = \langle\Phi_i|H_0|\Phi_i\rangle + \langle\Phi_i|H'|\Phi_i\rangle + \langle\Phi_i|H'Q_iH'|\Phi_i\rangle + \cdots \tag{1.17}$$

The great advantage of MBPT formulation is that the individual orders of perturbation expansion (1.17) can be treated diagrammatically using the technique of the Feynman-like diagrams, which makes the theory rather transparent.

Let us compare now the Hamiltonians (1.9) and (1.7). We see that they are formally identical except that all quantities in the Hamiltonian (1.7) are χ-dependent. If we further ask

$$f(\chi)|P(\chi)\rangle = \varepsilon_P(\chi)|P(\chi)\rangle \tag{1.18}$$

we can split the Hamiltonian (1.7) into the Moller–Plesset form with

$$\epsilon_0(\chi) = \langle \Phi_0(\chi)|H(\chi)|\Phi_0(\chi)\rangle \tag{1.19}$$

$$H_0(\chi) = \sum_P \varepsilon_P(\chi) N[a_P^+(\chi)a_P(\chi)] \tag{1.20}$$

and

$$H'(\chi) = \tfrac{1}{2} \sum_{PQRS} \langle P(\chi)Q(\chi)|v(\chi)|R(\chi)S(\chi)\rangle N[a_P^+(\chi)a_Q^+(\chi)a_S(\chi)a_R(\chi)] \tag{1.21}$$

To get the eigenvalue $E_i(\chi)$ from equation (1.1) we can use the standard nondegenerate Rayleigh–Schrödinger perturbation theory expansion (1.17) in the form

$$\begin{aligned} E_i(\chi) = {} & \langle \Phi_i\chi)|H_0(\chi)|\Phi_i(\chi)\rangle + \langle \Phi_i(\chi)|H'|\Phi_i(\chi)\rangle \\ & + \langle \Phi_i(\chi)|H'(\chi)Q_i(\chi)H'(\chi)|\Phi_i(\chi)\rangle + \cdots \end{aligned} \tag{1.22}$$

To satisfy equations (1.18) can be accomplished by solving the CPHF equations.[16] To get the creation ($a_P^+(\chi)$) and annihilation ($a_Q(\chi)$) operators we shall use the technique of canonical transformations. In our derivation we shall study a case in which the external perturbation $\sum_i \chi w_i$ will control the displacement of an atomic nucleus in the molecular system. The theory can be easily generalized. Further, we shall work with the total vibrational–electronic (V–E) Hamiltonian. If the reader wants to follow the derivation only with the electronic Hamiltonian it will be sufficient in what follows to work only with the Hamiltonian H_A from our total V–E Hamiltonian. The quantization of the V–E Hamiltonian is done hierarchically in a way that we first work in crude representation; then, using the technique of canonical transformations, we work in adiabatic representation. The theory is presented in detail. Diagrammatic rules for the Feynman-like diagrams, which can be used in MBPT, are also developed. The total V–E Hamiltonian is then used to derive the method for the calculation of the energy of the first vibrational transitions in molecules based on MBPT. The present theory permits us to derive the explicit formulae for the energy of the first vibrational transition in molecules. The corrections to harmonic frequencies $\hbar\omega$ calculated on the self-consistent field (SCF) level are due to electron correlation, anharmonicity effects, and the nonadiabatic effects.[23,24]

2. The Molecular Vibrational–Electronic (V–E) Hamiltonian

2.1. Quantization of Molecular V–E Hamiltonian

In this section we shall address the quantization of the total molecular V–E Hamiltonian. Our aim is to bring this Hamiltonian into the form suitable for many-body perturbation theory calculations.

Our molecular V–E Hamiltonian is

$$H = H_{NN}(R) + H_{EN}(r, R) + H_{EE}(r) \tag{2.1}$$

where $H_{NN}(R)$ is the nuclei term

$$H_{NN}(R) = T_N(R) + E_{NN}(R) \tag{2.2}$$

representing the kinetic energy of the nuclei (T_N) and interaction between nuclei (E_{NN}), and R denotes the nuclear coordinates. The last two terms of the Hamiltonian in equation (2.1) correspond to the standard electronic Hamiltonian in equation (1.8)

$$H_{EN}(r, R) + H_{EE}(r) = h + v^0 \tag{2.3}$$

where h is the one-electron part representing the kinetic energy of the electrons and the electron–nuclear attraction term, v^0 is the two-electron part of the Hamiltonian corresponding to the electron–electron repulsion term, and r denotes the electron coordinates. For the purpose of diagrammatic many-body perturbation theory, it will be efficient to work in second quantization formalism.[4,7,9] The electronic Hamiltonian of equation (2.3) has the form

$$H_{EN} + H_{EE} = \sum_{PQ} \langle P|h|Q\rangle a_P^+ a_Q + \tfrac{1}{2} \sum_{PQRS} \langle PQ|v^0|RS\rangle a_P^+ a_Q^+ a_S a_R, \tag{2.4}$$

where $a_P^+(a_Q)$ is the creation (annihilation) operator for electrons in the spin orbital basis $|P\rangle, |Q\rangle, \ldots$. If we apply the Wick theorem to (2.4) we can rewrite (2.4) as

$$\begin{aligned} H_{EN} + H_{EE} = {} & \sum_I h_{II} + \tfrac{1}{2} \sum_{IJ} (v^0_{IJIJ} - v^0_{IJJI}) \\ & + \sum_{PQ} h_{PQ} N[a_P^+ a_Q] + \sum_{PQI} (v^0_{PIQI} - v^0_{PIIQ}) N[a_P^+ a_Q] \\ & + \tfrac{1}{2} \sum_{PQRS} v^0_{PQRS} N[a_P^+ a_Q^+ a_S a_R] \end{aligned} \tag{2.5}$$

where v^0_{ABAB} (v^0_{ABBA}) denotes the coulomb (exchange) integral. Note that the operators (2.2) and (2.5) depend on the nuclear coordinate R. The operators (2.2) and (2.5) together represent the complete molecular V–E Hamiltonian.

To proceed further we have to specify the one-particle basis set $|P\rangle$, $|Q\rangle, \ldots$ in which we define the electron creation and annihilation operators. One possibility is to use the fixed basis set in crude representation. In this approximation the spin orbital basis set $|P\rangle, |Q\rangle, \ldots$ is determined at some fixed (equilibrium) nuclear coordinate R_0.

Of course, using the fixed basis set does not describe properly the physical situation, because we need to have the R-dependent basis set in which the electrons will follow adiabatically the motion of nuclei. Using the formalism of second quantization we use the technique of canonical transformations. Therefore, we first quantize the V–E Hamiltonian in crude representation, and then through the canonical transformation we requantize it in adiabatic representation.

Let us fix the nuclear coordinate R at some point R_0. We divide the individual terms of the Hamiltonian in (2.1) into two parts:

1. Terms that are determined at point R_0
2. Terms that represent the shift with respect to point R_0. (We shall use a prime to denote these terms.)

We can rewrite the electronic part of the V–E Hamiltonian in equation (2.5) as

$$H_{EN} + H_{EE} = E^0_{SCF} + h'_{SCF} + \sum_P \varepsilon_P N[a^+_P a_P]$$

$$+ \sum_{PQ} h'_{PQ} N[a^+_P a_Q] + \tfrac{1}{2} \sum_{PQRS} v^0_{PQRS} N[a^+_P a^+_Q a_S a_R] \qquad (2.6)$$

where E^0_{SCF} is the Hartree–Fock energy calculated at the point R_0, and h'_{SCF} is the shift in the Hartree–Fock energy with respect to the point other than R_0. The same is true for the one-particle operator of (2.6), where ε_P are the one-particle Hartree–Fock energies calculated at point R_0. The correlation operator is not changed because it does not depend on nuclear coordinates R. Furthermore, we use the following notation

$$h'_{SCF} = u'_{SCF} \qquad (2.7)$$

$$h'_{PQ} = u'_{PQ} \qquad (2.8)$$

where

$$u = \sum_{ij} \frac{-Z_j e^2}{|\mathbf{r}_i - \mathbf{R}_j|} \tag{2.9}$$

$$u_{PQ} = \left\langle P \middle| \sum_j \frac{-Z_j e^2}{|\mathbf{r} - \mathbf{R}_j|} \middle| Q \right\rangle \tag{2.10}$$

and

$$u_{SCF} = \sum_I u_{II} \tag{2.11}$$

Let us perform the Taylor expansion for the energies E_{NN} and u_{SCF} about the point R_0:

$$E_{NN} = E_{NN}^{(0)} + E'_{NN} = \sum_{i=0}^{\infty} E_{NN}^{(i)} \tag{2.12}$$

and

$$u_{SCF} = u_{SCF}^{(0)} + u'_{SCF} = \sum_{i=0}^{\infty} u_{SCF}^{(i)} \tag{2.13}$$

Using (2.12) and (2.13) we can rewrite the Hamiltonian in (2.1) as

$$\begin{aligned} H = {} & E_{NN}^{(0)} + E_{SCF}^{(0)} + T_N + E_{NN}^{(2)} + u_{SCF}^{(2)} + \sum_P \varepsilon_P N[a_P^+ a_P] \\ & + \tfrac{1}{2} \sum_{PQRS} \overset{0}{v}_{PQRS} N[a_P^+ a_Q^+ a_S a_R] + E'_{NN} - E_{NN}^{(2)} \\ & + u'_{SCF} - u_{SCF}^{(2)} + \sum_{PQ} u'_{PQ} N[a_P^+ a_Q] \end{aligned} \tag{2.14}$$

Nothing is changed in the Hamiltonian of equation (2.1), and therefore the Hamiltonian in (2.14) represents the complete V–E Hamiltonian (2.1).

We can now define the harmonic oscillators through which we define the boson (phonon) creation (b^+) and boson (phonon) annihilation (b) operators

$$T_N + E_{NN}^{(2)} + u_{SCF}^{(2)} = \sum_r \hbar\omega_r (b_r^+ b_r + 1/2) \tag{2.15}$$

where ω_r is the frequency of the harmonic oscillator.

This is an important step but we note that it is not the only possibility to define the harmonic oscillators in the V–E Hamiltonian. Our aim is to split the V–E Hamiltonian (2.14) into the unperturbed part and the perturbation, and the natural separation will be to take the harmonic oscillators as the unperturbed part of the Hamiltonian. Defining the harmonic oscillators in different ways gives us different perturbations.

Using (2.15) we can rewrite the Hamiltonian in (2.14) as

$$H = E_{NN}^{(0)} + E_{SCF}^{(0)} + \sum_r \hbar\omega_r(b_r^+ b_r + 1/2) + \sum_P \varepsilon_P N[a_P^+ a_P] + \tfrac{1}{2} \sum_{PQRS} v_{PQRS}^0 N[a_P^+ a_Q^+ a_S a_R] + E'_{NN} - E_{NN}^{(2)} + u'_{SCF} - u_{SCF}^{(2)} + \sum_{PQ} u'_{PQ} N[a_P^+ a_Q] \tag{2.16}$$

To use the perturbation theory we have to split the Hamiltonian in (2.16) into the unperturbed part (H_0) and the perturbation (H')

$$H = H_0 + H' \tag{2.17}$$

Due to the crude approximation, we can partition the Hamiltonian in (2.16) as

$$H_0 = E_{NN}^{(0)} + E_{SCF}^{(0)} + \sum_P \varepsilon_P N[a_P^+ a_P] + \sum_r \hbar\omega_r(b_r^+ b_r + 1/2) \tag{2.18}$$

and

$$H' = H'_E + H'_F + H'_I \tag{2.19}$$

We do not specify explicitly the terms in (2.19); this will be done later. But let H'_E correspond to the electron correlation operator, H'_F represent the vibrational operator, and H'_I represent the electronic–vibrational operator.

In equation (2.16) all quantities were defined through Cartesian coordinates. For further purposes it will be natural to work in normal coordinates $\{Q_r\}$. The normal coordinate in second quantized formalism is given as

$$Q_r \sim (b_r + b_r^+) \tag{2.20}$$

Therefore, to work with the Hamiltonian in (2.16) and to use the Taylor expansion in (2.12) and (2.13) in normal coordinate space, we shall have to work with products of $(b_r + b_r^+)$ operators.

Let us first analyze the product of two normal coordinates

$$Q_r \cdot Q_s \approx (b_r + b_r^+) \cdot (b_s + b_s^+) \tag{2.21}$$

and then we generalize it to any number of Q_r operators in order to describe any order of Taylor expansion in normal coordinates space. According to the Wick theorem we have

$$\begin{aligned}(b_r + b_r^+) \cdot (b_s + b_s^+) = {} & N[(b_r + b_r^+) \cdot (b_s + b_s^+)] + N[(\underline{br + b_r^+) \cdot (b_s} + b_s^+)] \\ & + N[(\underline{b_r + b_r^+) \cdot (b_s + b_s^+})] + N[(b_r + \underline{b_r^+) \cdot (b_s} + b_s^+)] \\ & + N[(b_r + \underline{b_r^+) \cdot (b_s + b_s^+})]\end{aligned} \tag{2.22}$$

Because for boson creation and annihilation operators we have the following commutation relations

$$[b_r, b_s^+] = \delta_{rs} \tag{2.23}$$

$$[b_r, b_s] = 0 \tag{2.24}$$

$$[b_r^+, b_s^+] = 0 \tag{2.25}$$

and the contraction of two operators is defined as

$$\underline{\mathbf{M}_k\mathbf{M}_l} = M_k M_l - N[M_k M_l] \tag{2.26}$$

we can write

$$Q_r Q_s = N[Q_r Q_s] + N[\underline{\mathbf{Q}_r\mathbf{Q}_s}] \tag{2.27}$$

To generalize (2.27), and therefore to develop the Wick theorem for products of $(b_r + b_r^+)$ operators, we proceed in the following way. Let us denote the vector boson operators as:

1. Vector boson annihilation operator

$$\mathbf{b} = \{b_r\} \tag{2.28}$$

2. Vector boson creation operator

$$\mathbf{b}^+ = \{b_r^+\} \tag{2.29}$$

Let us define the operator $\mathbf{B}$ as

$$\mathbf{B} = \mathbf{b} + \mathbf{b}^+ \tag{2.30}$$

Furthermore, we denote by $\mathbf{P}^{(n)}$ the tensor of the order n and by $\mathbf{Q}^{(m)}$ the tensor of the order m. The tensor product we denote as $\mathbf{P}^{(n)} \times \mathbf{Q}^{(m)}$, and the scalar product we denote as $\mathbf{P}^{(n)} \cdot \mathbf{Q}^{(m)}$. We also define

$$P^{(n)k} = \underbrace{P^{(n)} \times P^{(n)} \times \cdots \times P^{(n)}}_{k} \tag{2.31}$$

If we denote by $\mathbf{S}^{(n)}$ the totally symmetric tensor of the nth order, which stands for the partial derivatives with respect to the normal coordinate, then we can express generally the nth order of the Taylor expansion in normal coordinates space as

$$\begin{aligned}
\mathbf{S}^{(n)} \cdot \mathbf{B}^n = {} & \mathbf{S}^{(n)} \cdot \mathcal{N}[\mathbf{B}^n] + \binom{n}{2} \mathbf{S}^{(n)} \cdot \mathcal{N}[\underline{\mathbf{B} \times \mathbf{B}} \times \mathbf{B}^{n-2}] \\
& + \binom{n}{4} \frac{4!}{2!2^2} \mathbf{S}^{(n)} \cdot \mathcal{N}[(\underline{\mathbf{B} \times \mathbf{B}})^2 \times \mathbf{B}^{n-4}] \\
& + \cdots \begin{cases} + \dfrac{n!}{(n/2)!2^{n/2}} \mathbf{S}^{(n)} \cdot \mathcal{N}[(\underline{\mathbf{B} \times \mathbf{B}})^{n/2}] & \text{for } n \text{ even} \\[2ex] + \dfrac{n(n-1)!}{[(n-1)/2]!2^{(n-1)/2}} \mathbf{S}^{(n)} \cdot \mathcal{N}[(\underline{\mathbf{B} \times \mathbf{B}})^{(n-1)/2} \times \mathbf{B}] & \end{cases} \\
& \qquad \text{for } n \text{ odd}
\end{aligned} \tag{2.32}$$

Equation (2.32) can be further simplified as

$$\mathbf{S}^{(n)} \cdot \mathbf{B}^{(n)} = \sum_{k=0}^{[n/2]} \binom{n}{2k} \frac{(2k)!}{k!2^k} \mathbf{S}^{(n)} \cdot \mathcal{N}[(\underline{\mathbf{B} \times \mathbf{B}})^k \times \mathbf{B}^{n-2k}] \tag{2.33}$$

The quantity

$$\mathbf{B} \times \mathbf{B} = \mathbf{I}^{(2)} \tag{2.34}$$

is the identity tensor of the second order, and from $[n/2]$ we take only the integer part. Let us further introduce

$$\mathbf{B}^{(n)} = \{B_{r_1} \cdots r_n\} = \frac{1}{n!} N[\mathbf{B}^n] \tag{2.35}$$

We can finally write

$$\begin{aligned} \mathbf{S}^{(n)} \cdot \mathbf{B}^n &= \sum_{k=0}^{[n/2]} \frac{n!}{(2k)!(n-2k)!} \frac{(2k)!}{k!2^k} \mathbf{S}^{(n)} \cdot (N[\mathbf{B}^{n-2k}] \times \mathbf{I}^{(2)k}) \\ &= n! \sum_{k=0}^{[n/2]} \frac{1}{k!2^k} \mathbf{S}^{(n)} \cdot (\mathbf{B}^{(n-2k)} \times \mathbf{I}^{(2)k}) \end{aligned} \tag{2.36}$$

We can further simplify equation (2.36) if we take into account the Taylor expansion of the function of classical normal coordinates. Let Q be the set of classical normal coordinates $\{Q_r\}$ and $A(Q)$ be the function of classical normal coordinates that we write as a Taylor expansion in the following way:

$$A(Q) = \sum_{n=0}^{\infty} A^{(n)}(Q) \tag{2.37}$$

If we now put $\mathbf{B} = \mathbf{Q}$ we can write in second quantization

$$A^{(n)}(Q) = \frac{1}{n!} \mathbf{A}^{(n)} \cdot \mathbf{B}^n \tag{2.38}$$

where

$$\mathbf{A}^{(n)} = \{A^{r_1 \cdots r_n}\} = \left\{ \frac{\partial^n A(Q)}{\partial Q_{r_1} \cdots \partial Q_{r_n}} \bigg|_{Q_{r_1} = \cdots = Q_{r_n} = 0} \right\} \tag{2.39}$$

Using (2.36), (2.37), and (2.38) we can write

$$\mathbf{A}^{(n)}(Q) = \mathbf{A}^{(n)} \cdot \sum_{k=0}^{[n/2]} \frac{1}{k!2^k} (\mathbf{B}^{(n-2k)} \times \mathbf{I}^{(2)k}) \tag{2.40}$$

If we define

$$\mathbf{A}^{(k,m)} = \{A^{\overbrace{00\cdots0}^{2k} r_1 r_2 \cdots r_m}\} = \left\{\frac{1}{k!2^k} \sum_{g_1 g_2 \cdots g_k} \mathbf{A}^{g_1 g_1 g_2 g_2 \cdots g_k g_k r_1 r_2 \cdots r_m}\right\} \tag{2.41}$$

we finally arrive at the expression

$$\mathbf{A}(Q) = \sum_{n=0}^{\infty} \sum_{k=0}^{[n/2]} \mathbf{A}^{(k,n-2k)} \cdot \mathbf{B}^{(n-2k)} \tag{2.42}$$

Using (2.42) we can finally write the Hamiltonian in (2.17) in explicit form, that is,

$$\begin{aligned} H = {} & E_{NN}^{(0)} + E_{SCF}^{(0)} + \sum_P \varepsilon_P N[a_P^+ a_P] + \sum_r \hbar\omega_r (b_r^+ b_r + 1/2) \\ & + H'_E\left\{ \equiv \tfrac{1}{2} \sum_{PQRS} v_{PQRS}^0 N[a_P^+ a_Q^+ a_S a_R] \right\} \\ & + H'_F\left\{ \equiv \sum_{\substack{n=1 \\ n \neq 2}}^{\infty} \sum_{k=0}^{[n/2]} (\mathbf{E}_{NN}^{(k,n-2k)} + \mathbf{u}_{SCF}^{(k,n-2k)}) \cdot \mathbf{B}^{(n-2k)} \right\} \\ & + H'_I\left\{ \equiv \sum_{n=1}^{\infty} \sum_{k=0}^{[n/2]} \sum_{PQ} \mathbf{u}_{PQ}^{(k,n-2k)} \cdot \mathbf{B}^{(n-2k)} N[a_P^+ a_Q] \right\} \end{aligned} \tag{2.43}$$

The term H'_E is the electron correlation operator, H'_F corresponds to phonon–phonon interaction, and H'_I corresponds to electron–phonon interaction.

Let us comment on this partitioning of the V–E Hamiltonian. Because we have used the crude approximation the electrons are fixed at R_0, that is, at the minimum of the electronic energy. Due to this fact this approximation does not describe properly the physical situation (the electrons do not feel the R dependence due to the nuclei term). The perturbation term H_I that corresponds to electron–phonon interaction is too large, and the perturbation theory based on this partitioning of the V–E Hamiltonian will not converge. Therefore, in the next section we shall study the partitioning of the

V–E Hamiltonian in which we couple together the electronic and vibrational motions and thus make the electrons feel the R dependence due to the presence of nuclei terms. As has already been noted, the Hamiltonian in (2.16) is presented in real nuclear coordinates and the Hamiltonian in (2.43) in normal coordinates in second quantization. The transition from real nuclear coordinates to classical normal coordinates is a standard matter. However, let us recall a few well-known formulas. We do this also to specify the notation.

2.2. Transition from Nuclear to Normal Coordinates

The solution of the system of coupled oscillators (which are represented by the oscillating nuclei) normally takes on the form

$$n_{i\alpha} = R_{i\alpha} - R_{0i\alpha} = A_{i\alpha}\cos(\omega t), \tag{2.44}$$

where deviations $n_{i\alpha}$ (i represents the ith nucleus and α represents a Cartesian coordinate) are determined from the equation of harmonic motion. Unknown amplitudes $A_{i\alpha}$ and all the possible values of oscillator frequencies of the system can be found by solving the equation

$$\sum_{j\beta} (E_{\text{pot}}^{ij\alpha\beta} - m_i\omega^2\,\delta_{ij}\,\delta_{\alpha\beta})A_{j\beta} = 0 \tag{2.45}$$

where the indices $ij\alpha\beta$ have the meaning of derivatives with respect to real coordinates, m_i stands for the mass of the ith nucleus, and for the potential energy E_{pot},

$$E_{\text{pot}} = E_{NN}^{(2)} + V_N^{(2)} \tag{2.46}$$

where the effective potential $V_N^{(2)}$ taken in crude representation is $u_{SCF}^{(2)}$ (2.15) and in Born–Oppenheimer representation is $E_{SCF}^{(2)}$, as will be seen later [see, e.g., equation (4.42)]. Further, we shall address the selection of the $V_N^{(2)}$ potential.

Solving equation (2.45) we obtain the set of solutions, ω_r, and their corresponding amplitudes, $A_{i\alpha}^r$. If we introduce the set of normal coordinates, Q_r, we get for the deviations $n_{i\alpha}$ in (2.44)

$$n_{i\alpha} = \sum_r A_{i\alpha}^r Q_r \tag{2.47}$$

Substituting (2.47) into the total oscillation energy of the system E_{osc} we obtain

$$E_{\text{osc}} = E_{\text{kin}} + E_{\text{pot}} = \sum_r \left(\frac{1}{2M_r} P_r^2 + \tfrac{1}{2} M_r \omega_r^2 Q_r^2 \right) \tag{2.48}$$

where P_r are generalized momenta dual to the normal coordinates, and M_r are corresponding generalized masses as given by the following equation:

$$M_r = \sum_{i\alpha} m_i (A^r_{i\alpha})^2 \tag{2.49}$$

Equation (2.48) represents the system of independent harmonic oscillators. The normal coordinate in second quantization can be written as

$$Q_r = \left(\frac{\hbar}{2M_r \omega_r} \right)^{1/2} (b_r + b_r^+) \tag{2.50}$$

If we further put

$$M_r = \frac{\hbar}{2\omega_r} \tag{2.51}$$

then equations (2.49) and (2.51) give us the normalization condition according to which we have

$$Q_r = (b_r + b_r^+) \tag{2.52}$$

Let us now consider a general function of real coordinates $F(R)$. For its kth-order partial derivative we introduce the notation

$$F^{i_1 \cdots i_k \alpha_1 \cdots \alpha_k} \equiv \left. \frac{\partial^k F(R)}{\partial R_{i_1 \alpha_1} \cdots \partial R_{i_k \alpha_k}} \right|_{R = R_0} \tag{2.53}$$

Consider also the same function F but as a function of normal coordinates $F(Q)$. Its partial derivatives [for notation see, e.g., equation (2.39)] can be expressed through the partial derivatives with respect to real coordinates according to

$$R^{r_1\cdots r_k} = \sum_{\substack{i_1\cdots i_k \\ \alpha_1\cdots\alpha_k}} F^{i_1\cdots i_k\alpha_1\cdots\alpha_k} A^{r_1}_{i_1\alpha_1} \cdots A^{r_k}_{i_k\alpha_k} \tag{2.54}$$

Let us further simplify the notation in the following way:

$$F^{i\alpha_1\cdots\alpha_k} = F^{\overbrace{i\cdots i}^{k}\alpha_1\cdots\alpha_k} \tag{2.55}$$

$$F^{ij\alpha_1\cdots\alpha_k} = F^{\overbrace{i\cdots i}^{k-1}j\alpha_1\cdots\alpha_k}, \qquad k \geq 2 \tag{2.56}$$

For practical calculations we first calculate the expressions u_{PQ} (2.10) and E_{NN} (2.2) in real coordinates, and then we have to transform them to normal coordinates. In the sense of equations (2.55) and (2.56) in the expression for u_{PQ} it is reasonable to consider only the coefficients $u_{PQ}^{i\alpha_1\cdots\alpha_k}$ and in the expression E_{NN} only the coefficients $E_{NN}^{i\alpha}$ for $k = 1$ and $E_{NN}^{ij\alpha_1\cdots\alpha_k}$ for $k \geq 2$.

According to (2.55) the transformation of the expression u_{PQ} will have the form

$$u_{PQ}^{r_1\cdots r_k} = \sum_{i\alpha_1\cdots\alpha_k} u_{PQ}^{i\alpha_1\cdots\alpha_k} A^{r_1}_{i\alpha_1} \cdots A^{r_k}_{i\alpha_k} \tag{2.57}$$

For the expression E_{NN} we can derive the following symmetry relations:

$$E_{NN}^{i\alpha_1\cdots\alpha_k} = -\sum_{j\neq i} E_{NN}^{ij\alpha_1\cdots\alpha_k}, \qquad k \geq 2 \tag{2.58}$$

$$E_{NN}^{ij\alpha_1\cdots\alpha_k} = (-1)^k E_{NN}^{ji\alpha_1\cdots\alpha_k} = (-1)^{l+1} E_{NN}^{\overbrace{i\cdots i}^{k-1}\overbrace{j\cdots j}^{l}\alpha_1\cdots\alpha_k},$$

$$1 \leq l \leq k-1 \quad \text{and} \quad k \geq 2 \tag{2.59}$$

$$E_{NN}^{ij\alpha_1\cdots\alpha_k} = E_{NN}^{ijP[\alpha_1\cdots\alpha_k]}, \qquad k \geq 2 \tag{2.60}$$

where $P[\alpha_1 \cdots \alpha_k]$ means an arbitrary permutation of indices $\alpha_1 \cdots \alpha_k$. Using (2.53), (2.58), and (2.60) we have finally the following expressions for E_{NN}

$$E_{NN}^{r} = \sum_{i\alpha} E_{NN}^{i\alpha} A^{r}_{i\alpha}, \qquad k = 1 \tag{2.61}$$

and for $k \geq 2$

$$E_{NN}^{r_1 \cdots r_k} = \frac{1}{2} \sum_{ij, i \neq j} \sum_{l=1}^{k-1} \sum_{\alpha_1 \cdots \alpha_k} E_{NN}^{\overbrace{i \cdots i}^{l} \overbrace{j \cdots j}^{k-l} \alpha_1 \cdots \alpha_k} \binom{k}{l}$$

$$A_{i\alpha_1}^{r_1} \cdots A_{i\alpha_l}^{r_l} A_{j\alpha_l+1}^{r_l+1} \cdots A_{j\alpha_k}^{r_k} + \sum_{i\alpha_1 \cdots \alpha_k} E_{NN}^{i\alpha_1 \cdots \alpha_k} A_{i\alpha_1}^{r_1} \cdots A_{i\alpha_k}^{r_k}$$

$$= \sum_{ij, i>j} \sum_{l=1}^{k-1} E_{NN}^{ij\alpha_1 \cdots \alpha_k} (-1)^{k-l+1} \binom{k}{l} A_{i\alpha_1}^{r_1} \cdots A_{i\alpha_l}^{r_l} A_{j\alpha_l+1}^{r_l+1} \cdots A_{j\alpha_k}^{r_k}$$

$$- \frac{1}{2} \sum_{ij, i \neq j} \sum_{\alpha_1 \cdots \alpha_k} E_{NN}^{ij\alpha_1 \cdots \alpha_k} [A_{i\alpha_1}^{r_1} \cdots A_{i\alpha_k}^{r_k} + (-1)^k A_{j\alpha_1}^{r_1} \cdots A_{j\alpha_k}^{r_k}]$$

$$= - \sum_{ij, i>j} \sum_{\alpha_1 \cdots \alpha_k} E_{NN}^{ij\alpha_1 \cdots \alpha_k} \prod_{l=1}^{k} (A_{i\alpha_l}^{r_l} - A_{j\alpha_l}^{r_l}) \tag{2.62}$$

3. Canonical Transformations Defined for the Mixed System of Fermions and Bosons

In this section we shall define the group of canonical transformations for creation $(a_P^+(\chi))$ and annihilation $(a_Q(\chi))$ operators. These transformations will couple together the electronic and vibrational motions. We shall study the properties of these transformations and we shall also compare our approach with that based on gradient techniques. Finally, we shall develop the Feynman-like diagrammatic technique which will make the theory rather transparent. For detailed understanding of techniques of canonical transformations we refer the reader to Ref. 25.

As we have said in the introduction, the term $\sum_i \chi w_i$ in the Hamiltonian of (1.2) will control the displacement of atomic nuclei. Therefore, we want to have the creation (annihilation) operators that create (annihilate) particles on the R-dependent basis set. So we have to pass from the basis set fixed at point $R_0\{|P\rangle\}$ to the generally R-dependent basis set $\{|P(R)\rangle\}$:

$$|P\rangle \rightarrow |P(R)\rangle = \sum_Q C_{QP} |Q\rangle \tag{3.1}$$

We define the new set of fermion second quantized operators $\{\bar{a}_P\}$, $\{\bar{a}_Q^+\}$ as well as the new set of boson second quantized operators $\{\bar{b}_r\}$, $\{\bar{b}_r^+\}$. Because in our transformation we use the mixed set of old second quantized operators, we look at this transformation as we do at the quasi-particle transformation,[25] and instead of electrons and phonons when speaking about the new set of second quantized operators, we speak about fermions and bosons.

Further, we ask that the operators of the new fermions commute with the new set of boson operators. This condition makes the formulation of MBPT for the V–E Hamiltonian relatively simple. We define the set of fermion annihilation operators through the old set of electron annihilation operators as

$$\bar{a}_P = a_P + \sum_Q \sum_{k=1}^{\infty} \frac{1}{k!} \sum_{r_1 \cdots r_k} C_{PQ}^{r_1 \cdots r_k} B_{r_1} \cdots B_{r_k} a_Q \tag{3.2}$$

where B_r are defined through (2.30) and therefore have the meaning of normal coordinate operators, and $C_{PQ}^{r_1 \cdots r_k}$ is the set of unknown coefficients that will be determined. Because it holds that

$$[B_r, B_s] = 0 \tag{3.3}$$

we can choose the coefficients $C_{PQ}^{r_1 \cdots r_k}$ as totally symmetric tensors in indices $r_1 \cdots r_k$. Therefore, we write

$$\frac{1}{k!} C_{PQ}^{r_1 \cdots r_k} B_{r_1} \cdots B_{r_k} = \frac{1}{k!} \mathbf{C}_{PQ}^{(k)} \cdot \mathbf{B}^k = C_{PQ}^{(k)} \tag{3.4}$$

where in the tensor $\mathbf{C}_{PQ}^{(k)}$ the symbol (k) means the set of k indices, $r_1, \ldots, r_k$ in which the tensor is totally symmetric, and in the scalar $C_{PQ}^{(k)}$ the symbol (k) means only the order of the Taylor expansion.

Transformation (3.2) can therefore be simplified as

$$\bar{a}_P = \sum_Q \sum_{k=0}^{\infty} \frac{1}{k!} \mathbf{C}_{PQ}^{(k)} \cdot \mathbf{B}^k a_Q = \sum_Q \sum_{k=0}^{\infty} C_{PQ}^{(k)} a_Q = \sum_Q C_{PQ} a_Q \tag{3.5}$$

where

$$C_{PQ} = \sum_{k=0}^{\infty} C_{PQ}^{(k)} = \delta_{PQ} + \sum_{k=1}^{\infty} C_{PQ}^{(k)} \tag{3.6}$$

The transformation for the new set of fermion creation operators $\bar{a}_P^+$ is given as a Hermitian conjugate to (3.2)

$$\bar{a}_P^+ = a_P^+ + \sum_Q \sum_{k=1}^{\infty} \frac{1}{k!} \sum_{r_1 \cdots r_k} C_{PQ}^{r_1 \cdots r_k *} B_{r_1} \cdots B_{r_k} a_Q^+ \tag{3.7}$$

or using (3.5)

$$\bar{a}_P^+ = \sum_Q \sum_{k=0}^{\infty} \frac{1}{k!} \mathbf{C}_{PQ}^{(k)*} \cdot \mathbf{B}^k a_Q^+ = \sum_Q \sum_{k=0}^{\infty} C_{PQ}^{(k)+} a_Q^+ = \sum_Q C_{PQ}^+ a_Q^+ \tag{3.8}$$

Note that we used

$$\mathbf{B} = \mathbf{B}^+ \tag{3.9}$$

Further, we have to guarantee that our new set of second quantized operators are fermions, that is, they satisfy the following anticommutation relations:

$$\begin{gathered} \{\bar{a}_P, \bar{a}_Q\} = 0, \qquad \{\bar{a}_P^+, \bar{a}_Q^+\} = 0 \\ \{\bar{a}_P^+, \bar{a}_Q\} = \delta_{PQ} \end{gathered} \tag{3.10}$$

Substituting (3.2) and (3.7) into (3.10) will give us some restrictions on the coefficients C_{PQ}.

If we denote the commutator by $[A, B]$ and use the commutator algebra

$$[A, BC] = [A, B]C + B[A, C] = \{A, B\}C - B\{A, C\} \tag{3.11}$$

$$\{A, BC\} = \{A, B\}C - B[A, C] = [A, B]C + B\{A, C\} \tag{3.12}$$

we find the following relations for the coefficients C_{PQ}:

$$\begin{gathered} [C_{PQ}, C_{RS}] = 0, \qquad [C_{PQ}, C_{RS}^+] = 0 \\ [C_{PQ}, a_R] = 0, \qquad [C_{PQ}, a_R^+] = 0 \end{gathered} \tag{3.13}$$

and the unitary condition

$$\sum_R C_{PR} C_{QR}^+ = \sum_R C_{PR}^+ C_{QR} = \sum_R C_{RP} C_{RQ}^+ = \sum_R C_{RP}^+ C_{RQ} = \delta_{PQ} \tag{3.14}$$

Because our new set of fermion second quantized operators satisfies the anticommutation rules and the transformation is unitary in (3.14), our transformation in (3.2) and (3.7) is the canonical transformation.

Because our aim is to quantize the whole V–E Hamiltonian in (2.16) we introduce the canonical transformation for the new set of boson second quantized operators

$$\bar{b}_r = b_r + \sum_{PQ} \sum_{k=0}^{\infty} \frac{1}{k!} \sum_{s_1 \cdots s_k} d_{rPQ}^{s_1 \cdots s_k} B_{s_1} \cdots B_{s_k} a_P^+ a_Q \tag{3.15}$$

where b_r is the old set of phonon annihilation operators, and $d_{rPQ}^{s_1 \cdots s_k}$ are new coefficients which have to be determined. We can choose them as a totally symmetric tensor in indices $s_1 \cdots s_k$. Then we can write

$$\frac{1}{k!} \sum_{s_1 \cdots s_k} d_{rPQ}^{s_1 \cdots s_k} B_{s_1} \cdots B_{s_k} = \frac{1}{k!} \mathbf{d}_{rPQ}^{(k)} \cdot \mathbf{B}^k = d_{rPQ}^{(k)} \tag{3.16}$$

Similarly, as in the case of (3.5) the transformation (3.15) can be simplified as

$$\begin{aligned} \bar{b}_r &= b_r + \sum_{PQ} \sum_{k=0}^{\infty} \frac{1}{k!} \mathbf{d}_{rPQ}^{(k)} \cdot \mathbf{B}^k a_P^+ a_Q \\ &= b_r + \sum_{PQ} \sum_{k=0}^{\infty} d_{rPQ}^{(k)} a_P^+ a_Q = b_r + \sum_{PQ} d_{rPQ} a_P^+ a_Q \end{aligned} \tag{3.17}$$

By making the Hermitian a conjugate we get the new set of boson creation operators

$$\bar{b}_r^+ = b_r^+ + \sum_{PQ} \sum_{k=0}^{\infty} \frac{1}{k!} \sum_{s_1 \cdots s_k} d_{rPQ}^{s_1 \cdots s_k *} B_{s_1} \cdots B_{s_k} a_Q^+ a_P \tag{3.18}$$

or

$$\bar{b}_r^+ = b_r^+ + \sum_{PQ} d_{rQP}^+ a_P^+ a_Q \tag{3.19}$$

For the new set of boson second quantized operators we ask that they satisfy the commutation relations

$$\begin{gathered} [\bar{b}_r, \bar{b}_s] = 0, \qquad [\bar{b}_r^+, \bar{b}_s^+] = 0 \\ [\bar{b}_r, \bar{b}_s^+] = \delta_{rs} \end{gathered} \tag{3.20}$$

These give us some restrictions on the coefficients d_{rPQ}. Further, we ask that our new set of fermion and boson second quantized operators commute, that is,

$$[\bar{a}_P, \bar{b}_r] = 0 \tag{3.21}$$

$$[\bar{a}_P, \bar{b}_r^+] = 0 \tag{3.22}$$

The conditions (3.21, 3.22) make the formulation of the many-body perturbation theory simple because we are able to split the V–E Hamiltonian in a way that the unperturbed wave function will be the product of the fermion and boson wave function. Further, conditions (3.21) and (3.22) couple together the C_{PQ} coefficients with d_{rPQ} coefficients in the following way:

$$[C_{PQ}, b_r] + \sum_R C_{PR} d_{rRQ} = 0 \tag{3.23}$$

Using the unitary condition (3.14) we get

$$d_{rPQ} = \sum_R C_{RP}^+ [b_r, C_{RQ}] \tag{3.24}$$

from which we get the important relation

$$d_{rPQ} = -d_{rQP}^+ \tag{3.25}$$

Similarly, using (3.22) we get

$$[C_{PQ}, b_r^+] + \sum_R C_{PR} d_{rQR}^+ = 0 \tag{3.26}$$

Finally, the relations (3.20) lead to

$$[d_{rPQ}, b_s] + [b_r, d_{sPQ}] + \sum_R (d_{rPR} d_{sRQ} - d_{rRQ} d_{sPR}) = 0 \tag{3.27}$$

and

$$[d_{rPQ}, b_s^+] + [b_r, d_{sQP}^+] + \sum_R (d_{rPR} d_{sQR}^+ - d_{rRQ} d_{sRP}^+) = 0 \tag{3.28}$$

If we introduce in analogy with (2.30)

$$\bar{B}_r = \bar{b}_r + \bar{b}_r^+ \tag{3.29}$$

and the total number operator

$$\bar{N} = \sum_P \bar{a}_P^+ \bar{a}_P \tag{3.30}$$

then substituting the canonical transformations (3.5), (3.8), (3.17), and (3.19) into (3.29) and (3.30) we find that we have two invariants of the transformation

$$\bar{\mathbf{B}} = \mathbf{B} \tag{3.31}$$

and

$$\bar{N} = N \tag{3.32}$$

This means that according to (3.31) by the canonical transformation we do not change the normal coordinate, and according to (3.32) we do not change the number of fermions.

The relation (3.32) means that we can redefine the Fermi vacuum. Instead of using the former electrons and phonons as building blocks of the molecule, we use the new building blocks, namely, fermions and bosons, where the former electrons and phonons are coupled together by the canonical transformation. As a further step we find the useful relations for the C_{PQ} and d_{rPQ} coefficients, which we need for the MBPT calculations. First, let us start with the unitary condition (3.14) and rewrite it in the recurrent form

$$\sum_R C_{RP}^+ C_{RQ} = \delta_{PQ} + C'_{PQ} + C'^+_{QP} + \sum_R C'^+_{RP} C'_{RQ} = \delta_{PQ} \tag{3.33}$$

where

$$C'_{PQ} = C_{PQ} - C_{PQ}^{(0)} = C_{PQ} - \delta_{PQ} \tag{3.34}$$

We denote the Taylor expansion for C'_{PQ} as

$$C'_{PQ} = \sum_{k=1}^{\infty} C_{PQ}^{(k)} \tag{3.35}$$

Then we get the following recurrent relations for the operators $C_{PQ}^{(k)}$:

$$C_{PQ}^{(1)} + C_{QP}^{(1)+} = 0 \tag{3.36}$$

$$C_{PQ}^{(k)} + C_{QP}^{(k)+} = -\sum_{l=1}^{k-1} \sum_{R} C_{RP}^{(l)+} C_{RQ}^{(k-l)}, \qquad k \geq 2 \tag{3.37}$$

In order to get the recurrent relations for coefficients $\mathbf{C}_{PQ}^{(k)}$ (3.4) we introduce the summation-permutation operator $P_{\alpha\beta}$ that acts on the product of two tensors $\mathbf{g}^{(k)}$ and $\mathbf{h}^{(k)}$ in the following way:

$$P_{\alpha\beta}\mathbf{g}^{(k)\alpha} \times \mathbf{h}^{(l)\beta} = \{P_{\alpha\beta}g^{(r_1\cdots r_k)\alpha}h^{(s_1\cdots s_l)\beta}\}$$

$$\equiv \left\{ \sum_{i=1}^{\binom{k}{l}} g^{(t_{(i)1}\cdots t_{(i)k})} h^{(t_{(i)k+1}\cdots t_{(i)k+l})} \right\} \tag{3.38}$$

where $t_{(i)1} \cdots t_{(i)k+l}$ is the ith permutation of the set $r_1 \cdots r_k\, s_1 \cdots s_l$, where we do not take into account the permutations in the subsets $r_1 \cdots r_k$ and $s_1 \cdots s_l$. Using (3.38) we get for (3.4)

$$\mathbf{C}_{PQ}^{(1)} + \mathbf{C}_{QP}^{(1)*} = 0 \tag{3.39}$$

$$\mathbf{C}_{PQ}^{(k)} + \mathbf{C}_{QP}^{(k)*} = -\sum_{l=1}^{k-1} \sum_{R} P_{\alpha\beta}\mathbf{C}_{RP}^{(l)\alpha*} \times \mathbf{C}_{RQ}^{(k-l)\beta}, \qquad k \geq 2 \tag{3.40}$$

For coefficients $\mathbf{d}_{rPQ}^{(r)}$ given by (3.24) we get the following expressions through coefficients $\mathbf{C}_{PQ}^{(r)}$:

$$d_{rPQ} = \sum_{k=0}^{\infty} \sum_{l=0}^{k} \frac{1}{l!(k-l)!} \sum_{R} (\mathbf{C}_{RP}^{(l)*} \times \mathbf{C}_{RQ}^{r(k-l)}) \cdot \mathbf{B}^k \tag{3.41}$$

Combining (3.41) and (3.16) we get

$$\mathbf{d}_{rPQ}^{(k)} = \sum_{l=0}^{k} \sum_{R} P_{\alpha\beta}\mathbf{C}_{RP}^{(l)\alpha*} \times \mathbf{C}_{RQ}^{r(k-l)\beta} \tag{3.42}$$

For the first and second orders we get for the coefficients $\mathbf{C}_{PQ}^{(k)}$

$$C^{r}_{PQ} + C^{r*}_{QP} = 0 \tag{3.43}$$

$$C^{rs}_{PQ} + C^{rs*}_{QP} = -\sum_{R} (C^{r*}_{RP}C^{s}_{RQ} + C^{s*}_{RP}C^{r}_{RQ}) \tag{3.44}$$

Because the coefficients $\mathbf{d}_{rPQ}^{(k)}$ can be expressed through the coefficients $\mathbf{C}_{PQ}^{(k)}$ we have

$$d_{rPQ}^{(0)} = C^{r}_{PQ} \tag{3.45}$$

$$d^{s}_{rPQ} = C^{rs}_{PQ} + \sum_{R} C^{s*}_{RP} C^{r}_{RQ} \tag{3.46}$$

We have presented the algebra of C_{PQ} and d_{rPQ} coefficients in detail because we have found it very important for further practical applications.

To close this section we would like to mention the important property of our canonical transformations, namely, that they form the mathematical group. Let us have the set of canonical transformations (3.2) for the annihilation operators

$$[a] = \{a_P, \bar{a}_P, \bar{\bar{a}}_P, \ldots\} \tag{3.47}$$

$$\bar{a}_P = \sum_{Q} C_{PQ} a_Q \tag{3.48}$$

and

$$\bar{\bar{a}}_P = \sum_{Q} \tilde{C}_{PQ} \bar{a}_Q \quad \text{and so on} \tag{3.49}$$

It is easy to show that the set of transformations (3.47) forms the mathematical group.[26] In a similar way we can proceed for fermion creation and boson creation and annihilation operators. This is an important property, because it guarantees the existence of an inverse transformation from the new set of fermions and bosons to the old set of electrons and phonons. This will be used in requantization of the crude representation of the V–E Hamiltonian.

4. Adiabatic Representation of the V–E Hamiltonian—Diagrammatic Technique

4.1. Requantization of the Total V–E Hamiltonian

The canonical transformation introduced in Section 3 permits us to requantize the total V–E Hamiltonian of (2.43). Instead of using the electrons and phonons as a building block, we use the new quasi-particles introduced by canonical transformations—fermions and bosons. A crucial step in requantization of the V–E Hamiltonian in adiabatic representation lies in choosing the $V_N^{(2)}$ potential.

Let us introduce the new effective potential $V_N^{(2)}(B)$, which will be specified later, through which we introduce the harmonic oscillators

$$H_{NN}^{(2)}(B) + V_N^{(2)}(B) = \sum_r \hbar\omega_r(b_r^+ b_r + 1/2) \tag{4.1}$$

By the dependence on (B) we specify that we work with normal coordinates. Let us divide the V–E Hamiltonian of (2.16) into two parts

$$H = H_A + B_B \tag{4.2}$$

where

$$H_A = E_{NN}(B) - E_{NN}^{(2)}(B) - V_N^{(2)}(B) + \sum_{PQ} h_{PQ}(B) a_P^+ a_Q + \tfrac{1}{2} \sum_{PQRS} v_{PQRS}^0 a_P^+ a_Q^+ a_S a_R \tag{4.3}$$

and

$$H_B = \sum_r \hbar\omega_r(b_r^+ b_r + 1/2) \tag{4.4}$$

The Hamiltonian of (4.2) with (4.3) and (4.4) is the total V–E Hamiltonian in crude representation. To requantize it we have to substitute for the electron creation (a_P^+) and annihilation (a_Q) operators as well as for the phonon creation (b_r^+) and annihilation (b_r) operators from our canonical transformations. Because the transformations form the group, there exist the inverse transformations.

Let us denote the inverse transformations as

$$a_P = \sum_Q \bar{C}_{PQ}(B) \bar{a}_Q \tag{4.5}$$

$$a_P^+ = \sum_Q \bar{C}_{PQ}(B) \bar{a}_Q^+ \tag{4.6}$$

$$b_r = \bar{b}_r + \sum_{PQ} \bar{d}_{rPQ}(B) \bar{a}_P^+ \bar{a}_Q \tag{4.7}$$

$$b_r^+ = \bar{b}_r^+ + \sum_{PQ} \bar{d}_{rQP}(B) \bar{a}_P^+ \bar{a}_Q \tag{4.8}$$

where a_P is the annihilation operator for the old set of electrons, and $\bar{a}_P$ is the annihilation operator for the new set of fermions, and so on. From the group properties of our transformations it follows that[23]

$$\bar{C}_{PQ} = C^{+}_{QP} \tag{4.9}$$

First let us requantize the operator H_A (4.3). Substituting (4.5) and (4.6) into (4.3) we have

$$H_A = E_{NN}(B) - E^{(2)}_{NN}(B) - V^{(2)}_{N}(B) + \sum_{PQRS} h_{RS}(B)\bar{C}_{RP}(B)\bar{C}_{SQ}(B)\bar{a}^{+}_{P}\bar{a}_{Q}$$
$$+ \tfrac{1}{2} \sum_{PQRSTUVW} v^{0}_{TUVW}\bar{C}_{TP}(B)\bar{C}_{UQ}(B)\bar{C}_{VR}(B)\bar{C}_{WS}(B)\bar{a}^{+}_{P}\bar{a}^{+}_{Q}\bar{a}_{S}\bar{a}_{R} \tag{4.10}$$

Because we found that the total number operator N (3.32) is the invariant of the transformation, we can redefine the vacuum state to be the Fermi vacuum of our new fermion quasi-particle operators. We use the indices I, J, K, L for occupied states; A, B, C, D for virtual states; and P, Q, R, S, T, U, V, W for either of them. To simplify the notation we further omit the writing of the dependence on the operator B, and we also omit the bar sign on top of the coefficients C and operators $a(a^{+})$. Using the Wick theorem[4] we can write the operator (4.10) in normal product form as

$$H_A = E_{NN} - E^{(2)}_{NN} - V^{(2)}_{N} + \sum_{RSI} h_{RS}C_{RI}C_{SI}$$
$$+ \tfrac{1}{2} \sum_{RSTUIJ} (v^{0}_{RTSU} - v^{0}_{RSTU})C_{RI}C_{SI}C_{TJ}C_{UJ}$$
$$+ \sum_{PQRS} h_{RS}C_{RP}C_{SQ}N[a^{+}_{P}a_{Q}]$$
$$+ \sum_{PQRSTUI} (v^{0}_{RTSU} - v^{0}_{RSTU})C_{RP}C_{SQ}C_{TI}C_{UI}N[a^{+}_{P}a_{Q}]$$
$$+ \sum_{PQRSTUVW} v^{0}_{TUVW}C_{TP}C_{UQ}C_{VR}C_{WS}N[a^{+}_{P}a^{+}_{Q}a_{S}a_{R}] \tag{4.11}$$

Using the concept of a quasi-particle Fermi vacuum we can define the new quasi-particle Hartree–Fock energy as

$$E_{SCF} = \sum_{RSI} h_{RS}C_{RI}C_{SI} + \tfrac{1}{2} \sum_{RSTUIJ} (v^{0}_{RTSU} - v^{0}_{RSTU})C_{RI}C_{SI}C_{TJ}C_{UJ} \tag{4.12}$$

the new Hartree–Fock operator f with the matrix elements

$$f_{PQ} = \sum_{RS} h_{RS} C_{RP} C_{SQ} + \sum_{RTSUI} (v^0_{RTSU} - v^0_{RSTU}) C_{RP} C_{SQ} C_{TI} C_{UI} \tag{4.13}$$

and the new two-particle integral

$$v_{PQRS} = \sum_{TUVW} v^0_{TUVW} C_{TP} C_{UQ} C_{VR} C_{WS} \tag{4.14}$$

This permits us to write the operator (4.10) in the form

$$H_A = E_{NN} - E^{(2)}_{NN} - V^{(2)}_N + E_{SCF} + \sum_{PQ} f_{PQ} N[a^+_P a_Q] + \tfrac{1}{2} \sum_{PQRS} v_{PQRS} N[a^+_P a^+_Q a_S a_R] \tag{4.15}$$

Except for the first three terms, which are due to the nuclei interaction, the rest of the operator H_A (4.15) is exactly the Moller–Plesset splitting[22] of the new quasi-particle fermion Hamiltonian. Note that the Hamiltonian of (4.15) has exactly the form (1.7) we discussed in the introduction. The quantities (4.12), (4.13), and (4.14) are analogous to (1.19), (1.18), and (1.21), respectively. To develop the many-body perturbation theory for the Hamiltonian H_A (4.15) we need to split it into the unperturbed part and the perturbation. Let us split the quantities E_{NN}, E_{SCF}, and f_{PQ} as

$$E_{NN} = E^0_{NN} + E'_{NN} \tag{4.16}$$

$$E_{SCF} = E^0_{SCF} + E'_{SCF} \tag{4.17}$$

$$f_{PQ} = f^0_{PQ} + f'_{PQ} \tag{4.18}$$

where the superscript zero corresponds to the equilibrium (or reference) internuclear distance.

We have

$$E^0_{SCF} = \sum_I h^0_{II} + \tfrac{1}{2} \sum_{IJ} (v^0_{IJIJ} - v^0_{IJJI}) \tag{4.19}$$

$$f^0_{PQ} = h^0_{PQ} + \sum_I (v^0_{PIQI} - v^0_{PQII}) = \epsilon_P \delta_{PQ} \tag{4.20}$$

Let us denote

$$h_{PQ} = h^0_{PQ} + u'_{PQ} \tag{4.21}$$

$$w'_{PQ} = u'_{PQ} + \sum_{RI} (2v^0_{PRQI} - v^0_{PRIQ} - v^0_{PIRQ})C'_{RI} \tag{4.22}$$

$$z'_{PQ} = \sum_{RS} (v^0_{PRQS} - v^0_{PRSQ})C'_{RI}C'_{SI} \tag{4.23}$$

Using (4.21)–(4.23) we get for E'_{SCF}

$$E'_{SCF} = \sum_{I} (u'_{II} + 2\epsilon_I C'_{II}) + \sum_{RI} (u'_{IR}C'_{RI} + w'_{IR}C'_{RI} + \epsilon_R C'_{RI}C'_{RI}) + \sum_{RSI} (w'_{RS}C'_{RI}C'_{SI} + \tfrac{1}{2}z'_{RS}C'_{RI}C'_{SI}) \tag{4.24}$$

Using (3.33) we have

$$C'_{II} = -\tfrac{1}{2} \sum_{R} C'_{RI}C'_{RI} \tag{4.25}$$

and therefore

$$E'_{SCF} = \sum_{I} u'_{II} + \sum_{RI} [(\epsilon_R - \epsilon_I)C'_{RI} + u'_{RI} + w'_{RI}]C'_{RI} + \sum_{RSI} (w'_{RS} + \tfrac{1}{2}z'_{RS})C'_{RI}C'_{SI} \tag{4.26}$$

Similarly, using (3.33) we get

$$f'_{PQ} = (\epsilon_P - \epsilon_Q)C'_{PQ} + \sum_{R} (\epsilon_R - \epsilon_Q)C'_{RP}C'_{RQ} + w'_{PQ} + z'_{PQ} + \sum_{R} [(w'_{RQ} + z'_{RQ})C'_{RP} + \text{h.c.}] + \sum_{RS} (w'_{RS} + z'_{RS})C'_{RP}C'_{SQ} \tag{4.27}$$

where h.c. stands for Hermitian conjugate. Now we are able to split the Hamiltonian H_A into the unperturbed part H^0_A and the perturbation H'_A

$$H_A = H^0_A + H'_A \tag{4.28}$$

$$H^0_A = E^0_{NN} + E^0_{SCF} + \sum_{P} \epsilon_P N[a^+_P a_P] \tag{4.29}$$

Let us denote for simplicity

$$E' = E'_{NN} - E^{(2)}_{NN} - V^{(2)}_{N} + E'_{SCF} \tag{4.30}$$

Then, using (4.26), (4.27), and (4.30) we can find for the perturbed part of the Hamiltonian in (4.28) the following three terms:

$$H'_{A^0} = E' \tag{4.31}$$

$$H'_{A'} = \sum_{PQ} f'_{PQ} N[a^+_P a_Q] \tag{4.32}$$

$$H'_{A''} = \tfrac{1}{2} \sum_{PQRS} v_{PQRS} N[a^+_P a^+_Q a_S a_R] \tag{4.33}$$

Because these three terms are given as the expansion through the normal coordinates we can use the expression (2.42) to obtain the final result:

$$H'_{A^0} = \sum_{n=1}^{\infty} H'_{A^0_n} = \sum_{n=1}^{\infty} \sum_{k=0}^{[n/2]} \mathbf{E}^{(k,n-2k)} \cdot \mathbf{B}^{(n-2k)} \tag{4.34}$$

$$H'_{A'} = \sum_{n=1}^{\infty} H'_{A'_n} = \sum_{n=1}^{\infty} \sum_{k=0}^{[n/2]} \sum_{PQ} \mathbf{f}^{(k,n-2k)}_{PQ} \cdot \mathbf{B}^{(n-2k)} N[a^+_P a_Q] \tag{4.35}$$

$$H'_{A''} = \sum_{n=0}^{\infty} H_{A''_n} = \tfrac{1}{2} \sum_{n=0}^{\infty} \sum_{k=0}^{[n/2]} \sum_{PQRS} \mathbf{v}^{(k,n-2k)}_{PQRS} \cdot \mathbf{B}^{(n-2k)} N[a^+_P a^+_Q a_S a_R] \tag{4.36}$$

Because the term (4.34) contains only boson N-product $B^{(n-2k)}$ this term corresponds to boson–boson interactions. The term in (4.35) contains the one-particle (fermion) N-product and boson N-product $B^{(n-2k)}$, and it represents the one-particle (fermion)–boson interactions. The last term (4.36) is due to two-particle boson interactions. These three terms, (4.34)–(4.36), can be represented diagrammatically using the Feynman-like diagrams as follows:

To the nth order of the term H'_{A^0} (4.34) we ascribe the following diagram

$$H_{A^0_n} = \sum_{k=0}^{[n/2]} \mathbf{E}^{(k,n-2k)} \cdot \mathbf{B}^{(n-2k)} \rightarrow \tag{4.37}$$

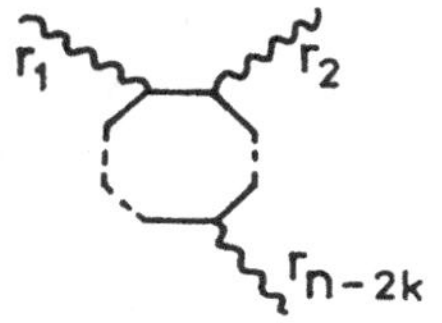

To the nth order of the term $\mathbf{H}'_{A'}$ in (4.35) we ascribe the diagram

$$\mathbf{H}'_{A'_n} = \sum_{k=0}^{[n/2]} \sum_{PQ} \mathbf{f}_{PQ}^{(k,n-2k)} \cdot \mathbf{B}^{(n-2k)} N[a_P^+ a_Q] \rightarrow \qquad (4.38)$$

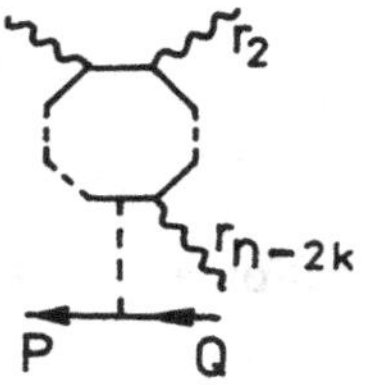

and to the nth order of $H'_{A''}$ in (4.36) the diagram

$$\mathbf{H}'_{A''_n} = \tfrac{1}{2} \sum_{n=0}^{[n/2]} \sum_{PQRS} \mathbf{v}_{PQRS}^{(k,n-2k)} \cdot \mathbf{B}^{(n-2k)} N[a_P^+ a_Q^+ a_S a_R]$$

$$\rightarrow \qquad (4.39)$$

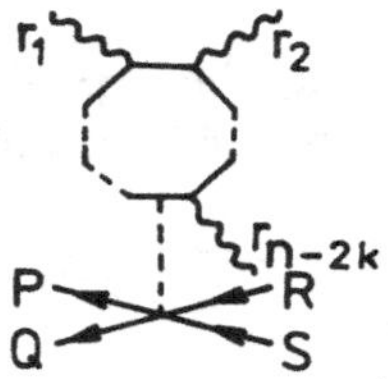

Because the nth order of individual perturbations is given through the polygon we use a convention such that the first order will be denoted by small circle "○", the second order by small abscissa "–", the third order by small triangle "△", etc. Thus, for example the diagram

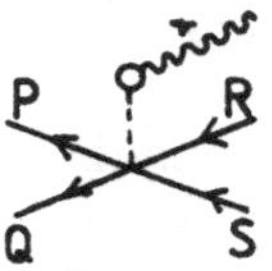

(4.40)

corresponds to the first derivative of the two-particle integral with respect to the normal coordinates. To make the perturbations as small as possible we ask the first order of the term $H'_{A_1^0}$ to be equal to zero, that is,

$$H'_{A_1^0} = E^{(1)} = E_{NN}^{(1)} + E_{SCF}^{(1)} = 0 \qquad (4.41)$$

This is nothing else than the condition that we optimize the geometry of the molecule on the SCF level.

We further ask the second order of $H'_{A_2^0}$ to be equal to zero, that is,

$$H'_{A_2^0} = E^{(2)} = E_{SCF}^{(2)} - V_N^{(2)} = 0 \qquad (4.42)$$

This means that we choose the force constant $V_N^{(2)}$ to be equal to the second derivative of the *SCF* energy with respect to the normal coordinates. Further, we try to simplify the perturbation $H'_{A'}$. This is achieved by making the terms $f_{PQ}^{(1)}, f_{PQ}^{(2)}, \ldots$ diagonal. For our practical applications it will be sufficient to diagonalize the term f_{PQ} to the first or to the second order of the expansion. This can be done by solving the *CPHF* equations.[14–16] We need to know the C_{PQ} coefficients to the first or to the second order. Their calculation will be shown later when discussing the connection of our approach with the gradient techniques. Further, we use the V–E Hamiltonian in (4.28) with (4.29)–(4.33) for the calculation of the first vibrational transition in some simple molecular system. To make these calculations feasible we have to limit ourselves to the finite orders of the expansions for the perturbations (4.34)–(4.36). We make such an approximation that the term H'_{A^0} is given up to the third order and the perturbations $H'_{A'}$ and $H'_{A''}$ up to the second order of the expansion. For this approximation we need to know the C_{PQ} coefficients up to the second order, and the perturbations (4.34)–(4.36) contain the terms

$$H'_{A^0} = H'_{A_1^0} + H'_{A_3^0} \tag{4.43}$$

$$H'_{A'} = H'_{A_1'} + H'_{A_2'} \tag{4.44}$$

$$H'_{A''} = H'_{A_0''} + H'_{A_1''} + H'_{A_2''} \tag{4.45}$$

Let us assume that we are able to diagonalize $f_{PQ}^{(1)}$ and $f_{PQ}^{(2)}$. Therefore, we have

$$f_{PQ}^{(1)} = \epsilon_P^{(1)}\,\delta_{PQ} \tag{4.46}$$

$$f_{PQ}^{(2)} = \epsilon_P^{(2)}\,\delta_{PQ} \tag{4.47}$$

We can form the following perturbations

$$H'_{A_1^0} = \sum_r E^r B_r \tag{4.48}$$

$$H'_{A_3^0} = \sum_{rst} E^{rst} B_{rst} + \sum_r E^{00r} B_r$$

$$H'_{A_1'} = \sum_{Pr} \epsilon_P^r B_r N[a_P^+ a_P] \tag{4.50}$$

$$H'_{A_2'} = \sum_{Prs} \epsilon_P^{rs} B_{Rs} N[a_P^+ a_P] + \sum_P \epsilon_P^{00} N[a_P^+ a_P] \tag{4.51}$$

$$H'_{A_0''} = \sum_{PQRS} v_{PQRS}^0 N[a_P^+ a_Q^+ a_S a_R] \tag{4.52}$$

$$H'_{A_1''} = \tfrac{1}{2} \sum_{PQRSr} v^{r}_{PQRS} B_r N[a_P^{+} a_Q^{+} a_S a_R] \tag{4.53}$$

$$H'_{A_2''} = \tfrac{1}{2} \sum_{PQRSrs} v^{rs}_{PQRS} B_{rs} N[a_P^{+} a_Q^{+} a_S a_R] + \tfrac{1}{2} \sum_{PQRS} v^{00}_{PQRS} N[a_P^{+} a_Q^{+} a_S a_R] \tag{4.54}$$

Using the diagrammatic convention (4.37)–(4.39) we have this diagrammatic representation of the perturbation H'_A.

$H'_{A_1^0} \rightarrow$  (4.55)

$H'_{A_3^0} \rightarrow$ 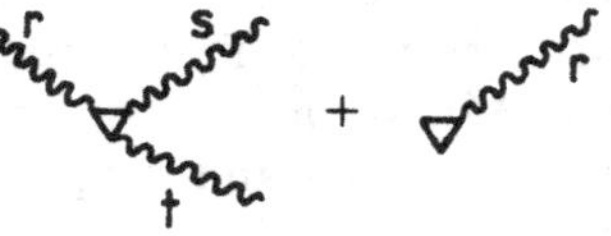 (4.56)

$H'_{A_1'} \rightarrow$ 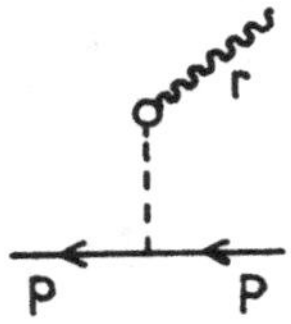 (4.57)

$H'_{A_2'} \rightarrow$ 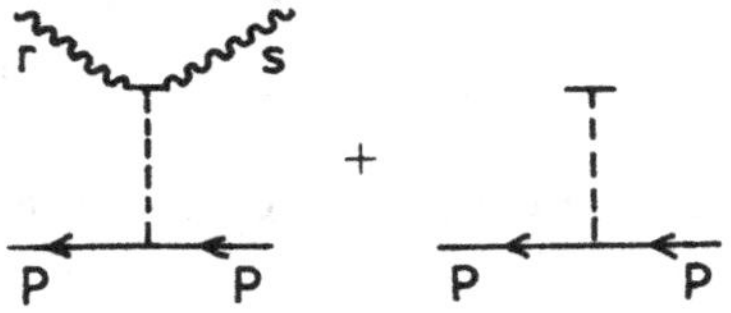 (4.58)

$H'_{A_0''} \rightarrow$  (4.59)

$H'_{A_1''} \rightarrow$ 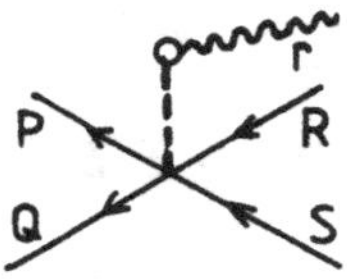 (4.60)

$H'_{A_2''} \rightarrow$ 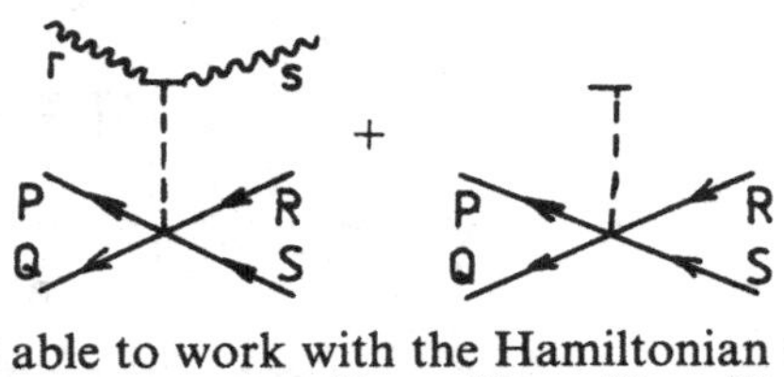 (4.61)

Note that in order to be able to work with the Hamiltonian we need to know the expressions for the quantities $E^{(1)}_{NN}$, $E^{(2)}_{NN}$, $E^{(3)}_{NN}$, $u^{(1)}_{PQ}$, $u^{(2)}_{PQ}$, and

$u_{PQ}^{(3)}$. We first calculate these quantities in real coordinates and then transform them to normal coordinates according to (2.57)–(2.62). Using (4.22) and (4.23) we get

$$w_{PQ}^{(1)} = u_{PQ}^{(1)} + \sum_{AI} (2v_{PAQI}^{0} - v_{PAIQ}^{0} - v_{PIAQ}^{0})C_{AI}^{(1)} \tag{4.62}$$

$$\begin{aligned} w_{PQ}^{(2)} = u_{PQ}^{(2)} &+ \sum_{AI} (2v_{PAQI}^{0} - v_{PAIQ}^{0} - v_{PIAQ}^{0})C_{AI}^{(2)} \\ &- \sum_{RIJ} (v_{PIQJ}^{0} - v_{PIJQ}^{0})C_{RI}^{(1)}C_{RJ}^{(1)} \end{aligned} \tag{4.63}$$

$$z_{PQ}^{(2)} = \sum_{RSI} (v_{PRQS}^{0} - v_{PRSQ}^{0})C_{RI}^{(1)}C_{SI}^{(1)} \tag{4.64}$$

Using (4.14) we get the following expressions for $v_{PQRS}^{(1)}$ and $v_{PQRS}^{(2)}$:

$$v_{PQRS}^{(1)} = \sum_{T} (v_{TQRS}^{0}C_{TP}^{(1)} + v_{PTRS}^{0}C_{TQ}^{(1)} + v_{PQTS}^{0}C_{TR}^{(1)} + v_{PQRT}^{0}C_{TS}^{(1)}) \tag{4.65}$$

$$\begin{aligned} v_{PQRS}^{(2)} = &\sum_{T} (v_{TQRS}^{0}C_{TP}^{(2)} + v_{PTRS}^{0}C_{TQ}^{(2)} + v_{PQTS}^{0}C_{TR}^{(2)} + v_{PQRT}^{0}C_{TS}^{(2)}) \\ &+ \sum_{TU} (v_{TURS}^{0}C_{TP}^{(1)}C_{UQ}^{(1)} + v_{TQUS}^{0}C_{TP}^{(1)}C_{UR}^{(1)} + v_{TQRU}^{0}C_{TP}^{(1)}C_{US}^{(1)} \\ &+ v_{PTUS}^{0}C_{TQ}^{(1)}C_{UR}^{(1)} + v_{PTRU}^{0}C_{TQ}^{(1)}C_{US}^{(1)} + v_{PQTU}^{0}C_{TR}^{(1)}c_{US}^{(1)}) \end{aligned} \tag{4.66}$$

Using (4.27) and diagonalizing $f_{PQ}^{(1)}$ we get $\epsilon_{P}^{(1)}$ and

$$f_{PQ}^{(1)} = (\epsilon_P - \epsilon_Q)C_{PQ}^{(1)} + w_{PQ}^{(1)} = \epsilon_{P}^{(1)}\,\delta_{PQ} \tag{4.67}$$

Equation (4.67) can be used to develop the iterative procedure to determine the coefficients $C_{PQ}^{(1)}$.[23] But instead of doing so, we show in Section 5 how to determine them through the *CPHF* equations.[14–16] Similarly, using (4.27) we get for $f_{PQ}^{(2)}$ the expression (4.68). Diagonalizing, we get $\epsilon_{P}^{(2)}$

$$\begin{aligned} f_{PQ}^{(2)} = (\epsilon_P - \epsilon_Q)C_{PQ}^{(2)} &+ \sum_{R} (\epsilon_R - \epsilon_Q)C_{RP}^{(1)}C_{RQ}^{(1)} \\ &+ (\epsilon_{P}^{(1)} - \epsilon_{Q}^{(1)})C_{PQ}^{(1)} + w_{PQ}^{(2)} + z_{PQ}^{(2)} = \epsilon_{P}^{(2)}\,\delta_{PQ} \end{aligned} \tag{4.68}$$

Finally, using (4.26) we get

$$E_{SCF}^{(1)} = \sum_I u_{II}^{(1)} \tag{4.69}$$

$$E_{SCF}^{(2)} = \sum_I u_{II}^{(2)} + \sum_{AI} u_{AI}^{(1)} C_{AI}^{(1)} \tag{4.70}$$

$$E_{SCF}^{(3)} = \sum_I (u_{II}^{(3)} - \epsilon_I^{(1)} C_{II}^{(2)}) + \sum_{AI} (u_{AI}^{(2)} - z_{AI}^{(2)}) C_{AI}^{(1)} + \sum_{RI} u_{RI}^{(1)} C_{RI}^{(2)} \tag{4.71}$$

So far we requantized only the H_A part of the V–E Hamiltonian in (4.2). We can proceed in an analogous way for the H_B part. From (4.4) we have

$$H_B = \sum_r \hbar\omega_r (b_r^+ b_r + 1/2) \tag{4.72}$$

Substituting the inverse transformations (4.7) and (4.8) into (4.72) we get

$$\begin{aligned} H_B = \sum_r \hbar\omega_r (\bar{b}_r^+ \bar{b}_r + 1/2) + \sum_{PQr} \hbar\omega_r (\bar{b}_r^+ \bar{d}_{rPQ} + \bar{d}_{rQP} \bar{b}_r) \\ + \sum_{PQRSr} \hbar\omega_r \bar{d}_{rQP} \bar{d}_{rRS} \bar{a}_P^+ \bar{a}_Q \bar{a}_R^+ \bar{a}_S \end{aligned} \tag{4.73}$$

Further, we omit the sign "–" on $\bar{d}_{rPQ}$ coefficients as well as on the $\bar{b}_r^+(\bar{b}_r)$ and $\bar{a}_P^+(\bar{a}_P)$ operators. Let us introduce the following functions of spin orbitals:

$$p(A) = 1, \qquad p(I) = 0, \qquad h(A) = 0, \qquad h(I) = 1 \tag{4.74}$$

Now we apply the Wick theorem[4] to (4.73) and get

$$a_P^+ a_Q = N[a_P^+ a_Q] + h(P)\, \delta_{PQ} \tag{4.75}$$

$$\begin{aligned} a_P^+ a_Q a_R^+ a_S = -N[a_P^+ a_R^+ a_Q a_S] + N[a_R^+ a_S] h(P)\, \delta_{PQ} - N[a_R^+ a_Q] h(P)\, \delta_{PS} \\ + N[a_P^+ a_S] p(Q)\, \delta_{QR} + N[a_P^+ a_Q] h(R)\, \delta_{RS} \\ + h(P) h(R)\, \delta_{PQ}\, \delta_{RS} + h(P) p(Q)\, \delta_{PS}\, \delta_{QR} \end{aligned} \tag{4.76}$$

Using (3.25), (4.75), and (4.76) we get

$$H_B = \sum_r \hbar\omega_r(b_r^+ b_r + 1/2) + \sum_{PQr} \hbar\omega_r(b_r^+ d_{rPQ} + d_{rQP} b_R) N[a_P^+ a_Q]$$
$$+ \sum_{AIr} \hbar\omega_r (d_{rAI})^2 + \sum_{PQAIr} \hbar\omega_r(d_{rPA} d_{rQA} - d_{rPI} d_{rQI}) N[a_P^+ a_Q]$$
$$+ \sum_{PQRSr} \hbar\omega_r d_{rPS} d_{rQR} N[a_P^+ a_Q^+ a_S a_R] \tag{4.77}$$

This is the complete H_B Hamiltonian requantized with respect to the new quasi-particles given by the transformations in (3.15) and (3.18). We can split it into the unperturbed part H_B^0 and the perturbations H'_B. The simplest way is to choose

$$H_B^0 = \sum_r \hbar\omega_r(b_r^+ b_r + 1/2) \tag{4.78}$$

and

$$H'_B = H'_{B^*} + H'_{B^0} + H'_{B'} + H'_{B''} \tag{4.79}$$

where

$$H'_{B^*} = \sum_{PQr} \hbar\omega_r(b_r^+ d_{rPQ} + d_{rQP}\, b_r) N[a_P^+ a_Q] \tag{4.80}$$

$$H'_{B^0} = \sum_{AIr} \hbar\omega_r (d_{rAI})^2 \tag{4.81}$$

$$H'_{B'} = \sum_{PQAIr} \hbar\omega_r(d_{rPA} d_{rQA} - d_{rPI} d_{rQI}) N[a_P^+ a_Q] \tag{4.82}$$

and

$$H'_{B''} = \sum_{PQRSr} \hbar\omega_r d_{rPS} d_{rQR} N[a_P^+ a_Q^+ a_S a_R] \tag{4.83}$$

Note that because the coefficients d_{rPQ} are given as an expansion, the perturbations (4.80)–(4.83) are also given as an expansion. To distinguish the index r in the coefficients d_{rPQ}—and therefore also in (4.80)–(4.83)—from the one that originates from the Taylor expansion according to the normal coordinates (e.g., $C_{PQ}^{r_1 \cdots r_k}$) we distinguish the former index r from the latter by prime. Therefore, we introduce the following notation

$$m'^{r}_{PQ} = \hbar\omega_r d_{rPQ} \tag{4.84}$$

and

$$\mathbf{B}^{(n),r} = b_r^+ \mathbf{B}^{(n)} - \mathbf{B}^{(n)} b_r \tag{4.85}$$

which gives us this expression for the perturbation H'_{B^*}:

$$\begin{aligned} H'_{B^*} = \sum_{n=0}^{\infty} H'_{B^*_{n+1}} &= \sum_{n=0}^{\infty} \sum_{k=0}^{[n/2]} \sum_{PQr} (\mathbf{m}_{PQ}^{(k,n-2k),r} \cdot b_r^+ \mathbf{B}^{(n-2k)} \\ &\quad + \mathbf{m}_{QP}^{(k,n-2k),r} \cdot \mathbf{B}^{(n-2k)} b_r) N[a_P^+ a_Q] \\ &= \sum_{n=0}^{\infty} \sum_{k=0}^{[n/2]} \sum_{PQr} \mathbf{m}_{PQ}^{(k,n-2k),r} \cdot \mathbf{B}^{(n-2k),r} N[a_P^+ a_Q] \end{aligned} \tag{4.86}$$

We can introduce the following diagrammatic representation. For the nth order of H'_{B^*} we have

$$\sum_{k=0}^{[n/2]} \sum_{PQr} (\mathbf{m}_{PQ}^{(k,n-2k),r} \cdot b_r^+ \mathbf{B}^{(n-2k)} + \mathbf{m}_{QP}^{(k,n-2k),r} \mathbf{B}^{(n-2k)} b_r) N[a_P^+ a_Q]$$

$\rightarrow$ 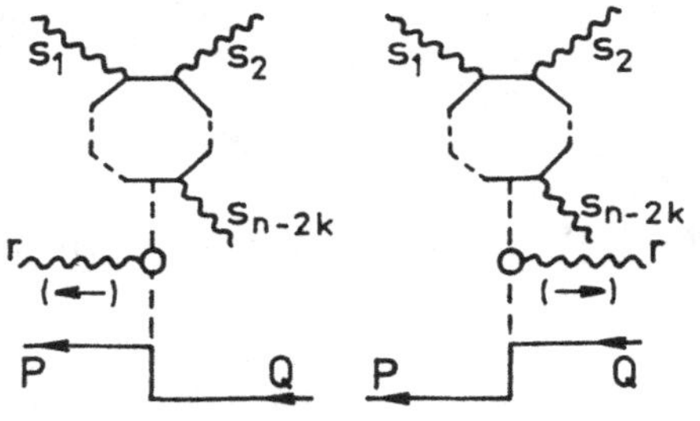 (4.87)

To simplify the notation in (4.82)–(4.83), where we have one-particle (fermion) and two-particle fermion operators, we introduce the following notation:

$$\tilde{f}'^{00}_{PQ} = \sum_{AIr} \hbar\omega_r (d_{rPA} d_{rQA} - d_{rPI} d_{rQI}) \tag{4.88}$$

$$\tilde{V}'^{00}_{PQRS} = 2 \sum_{r} \hbar\omega_r d_{rPS} d_{rQR} \tag{4.89}$$

and

$$\tilde{E}'^{00} = \sum_{AIr} \hbar\omega_r (d_{rAI})^2 \tag{4.90}$$

Now we write the expansions for the perturbations H'_{B^0}, $H'_{B'}$, and $H'_{B''}$ in the form

$$H'_{B^0} = \sum_{n=0}^{\infty} H'_{B^0_{n+2}} = \sum_{n=0}^{\infty} \sum_{k=0}^{[n/2]} \tilde{\mathbf{E}}^{(k,n-2k),00} \cdot \mathbf{B}^{(n-2k)} \tag{4.91}$$

$$H'_{B'} = \sum_{n=0}^{\infty} H'_{B'_n+2} = \sum_{n=0}^{\infty} \sum_{k=0}^{[n/2]} \sum_{PQ} \tilde{\mathbf{f}}^{(k,n-2k),00} \cdot \mathbf{B}^{(n-2k)} N[a_P^+ a_Q] \tag{4.92}$$

$$H'_{B''} = \sum_{n=0}^{\infty} H'_{B''_n+2} = \frac{1}{2} \sum_{n=0}^{\infty} \sum_{k=0}^{[n/2]} \sum_{PQRS} \tilde{\mathbf{v}}^{(k,n-2k),00} \cdot \mathbf{B}^{(n-2k)} N[a_P^+ a_Q^+ a_S a_R] \tag{4.93}$$

The corresponding nth order of these perturbations we represent diagrammatically as

$$\sum_{k=0}^{[n/2]} \tilde{\mathbf{E}}^{(k,n-2k),00} \cdot \mathbf{B}^{(n-2k)} \rightarrow \tag{4.94}$$

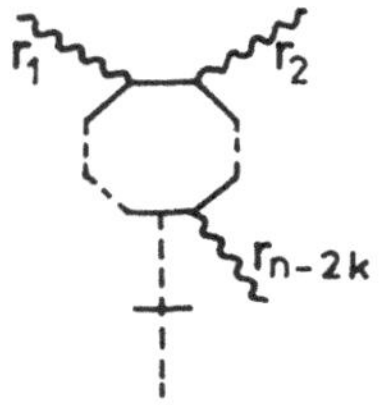

$$\sum_{k=0}^{[n/2]} \sum_{PQ} \tilde{\mathbf{f}}_{PQ}^{(k,n-2k),00} \cdot \mathbf{B}^{(n-2k)} N[a_P^+ a_Q] \rightarrow \tag{4.95}$$

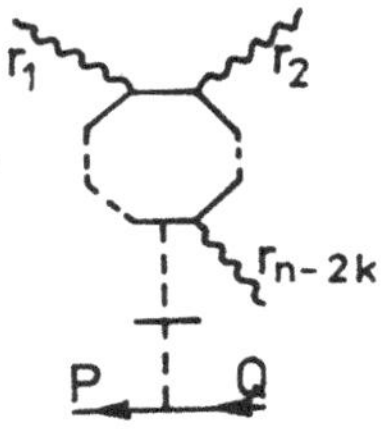

$$\frac{1}{2} \sum_{k=0}^{[n/2]} \sum_{PQRS} \tilde{\mathbf{v}}_{PQRS}^{(k,n-2k),00} \cdot \mathbf{B}^{(n-2k)} N[a_P^+ a_Q^+ a_S a_R] \rightarrow \tag{4.96}$$

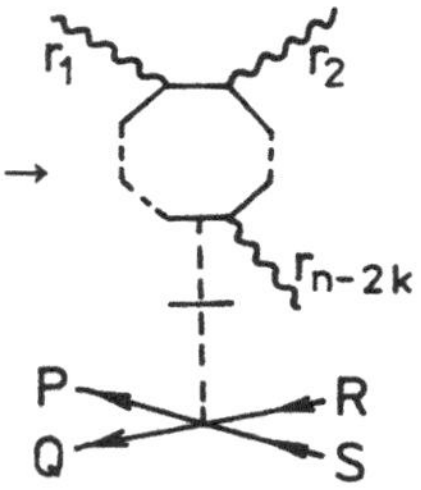

For practical reasons we limit ourselves to the final orders of the expansions (4.86) and (4.91)–(4.93). Let us again limit ourselves to the third order of expansion for the perturbation H'_{B^0} and to the second order for the perturbations H'_{B^*}, $H'_{B'}$, and $H'_{B''}$. Therefore we have

$$H'_{B^*} = H'_{B_1^*} + H'_{B_2^*} \tag{4.97}$$

$$H'_{B^0} = H'_{B_2^0} + H'_{B_3^0} \tag{4.98}$$

$$H'_{B'} = H'_{B_2'} \tag{4.99}$$

$$H'_{B''} = H'_{B_2''} \tag{4.100}$$

which is equivalent to the diagrams

$$H'_{B_1^*} = \sum_{PQr} m_{PQ}^{(0),r} B_{,r} N[a_P^+ a_Q] \rightarrow \tag{4.101}$$

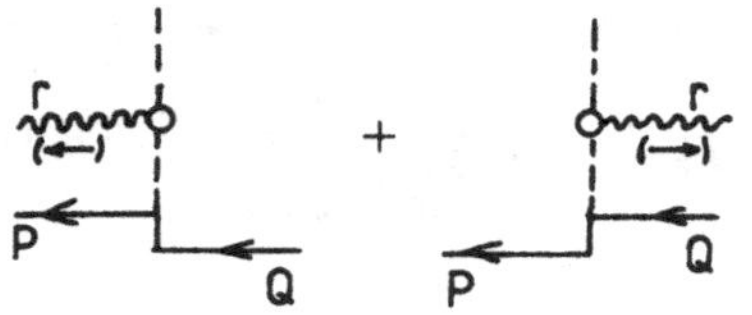

$$H'_{B_2^*} = \sum_{PQrs} m_{PQ}^{(s),r} B_{s,r} N[a_P^+ a_Q] \rightarrow \tag{4.102}$$

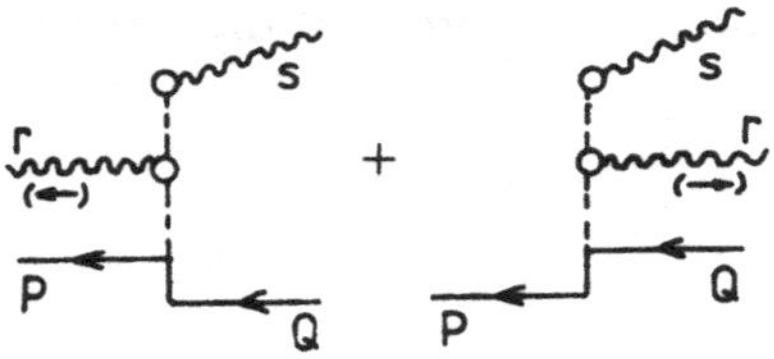

$$H'_{B_2^0} = \tilde{E}^{(0),00} \rightarrow \tag{4.103}$$

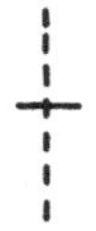

$$H'_{B_3^0} = \sum_r \tilde{E}^{(r),00} B_r \rightarrow \tag{4.104}$$

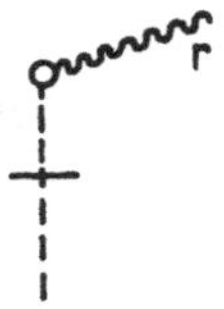

$$H'_{B_2'} = \sum_{PQ} \tilde{f}_{PQ}^{(0),00} N[a_P^+ a_Q] \rightarrow \tag{4.105}$$

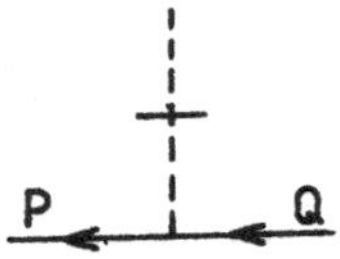

$$H'_{B_2''} = \frac{1}{2} \sum_{PQRS} \tilde{v}_{PQRS}^{(0),00} N[a_P^+ a_Q^+ a_S a_R] \rightarrow \tag{4.106}$$

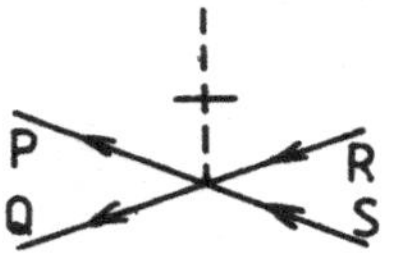

We present here the V–E Hamiltonian which we shall use in practical calculations:

$$H = \underbrace{E_{NN}^{0} + E_{SCF}^{0} + \sum_{P} \epsilon_{P} N[a_{P}^{+} a_{P}] + \sum_{r} \hbar\omega_{r}(b_{r}^{+} b_{r} + 1/2)}_{\text{unperturbed part}}$$

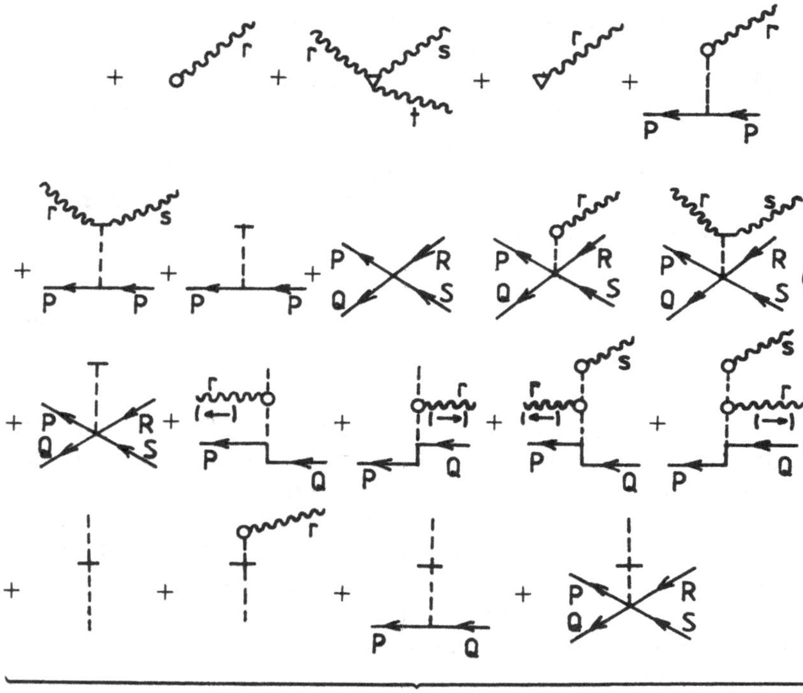

(4.107)

perturbations

Note that the only approximation in the Hamiltonian in (4.107) is due to the truncation of the expansion of the perturbations.

4.2. The Diagrammatic Rules

To use the diagrams in the V–E Hamiltonian in (4.107) for a many-body perturbation theory treatment we present the diagrammatic rules. The most general diagrammatic structure we have in perturbation theory expansion is the mixed fermion–boson diagram of the form

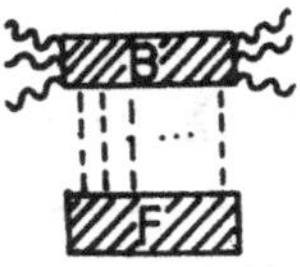

(4.108)

where $B(F)$ corresponds to bosons (fermions and the interrupted lines correspond to fermion–boson interaction). As a first step we present the topological or weight factor for such a diagram. To calculate the weight factor of the diagram in (4.108) we first separate the fermion and the boson part of the diagram, and therefore we calculate the weight factor separately for the boson part and the fermion part of the diagram. The total weight factor of the diagram in (4.108) is given as a product of the boson weight factor, W_B, and the fermion weight factor, W_F:

$$W = W_F W_B \tag{4.109}$$

The weight factor and the diagrammatic rules for the fermion diagrams are presented in many works.[4,7,9,11] Therefore, we present here only the derivation of the weight factor for the boson diagrams. Let us denote the weight factor of the vacuum diagram as $W_B(0)$. To make the derivation clear we introduce the notation we use. We call the polygon part in the skeletons (4.37)–(4.39), (4.87), and (4.96) the boson interaction vertex. This boson interaction vertex contains the vertices which are on the top of the skeleton, which we call simply boson coordinate vertices. There is still another type of vertex in the skeleton of (4.87) with the boson line in the middle of the skeleton, which we distinguish from the boson coordinate vertex. Because the vertex in the middle of the skeleton belongs to one boson vortex it does not contribute to the weight factor $W_B(0)$ because this is always multiplied by one. Therefore, the weight factor $W_B(0)$ is fully determined by the boson coordinate vertices and their interconnection with the boson lines. The weight factor $W_B(0)$ is given as a product of two terms,

$$W_B(0) = W_B(I)\, W_B(II) \tag{4.110}$$

The term W_B^I can be understood as follows. The normal product $N[B^n]$ can be represented diagrammatically as

$$\mathcal{N}[\mathbf{B}^n] = \{\mathcal{N}[(b_{r_1}^+ + b_{r_1})(b_{r_2}^+ + b_{r_2}) \cdots (b_{r_n}^+ + b_{r_n})]\}$$

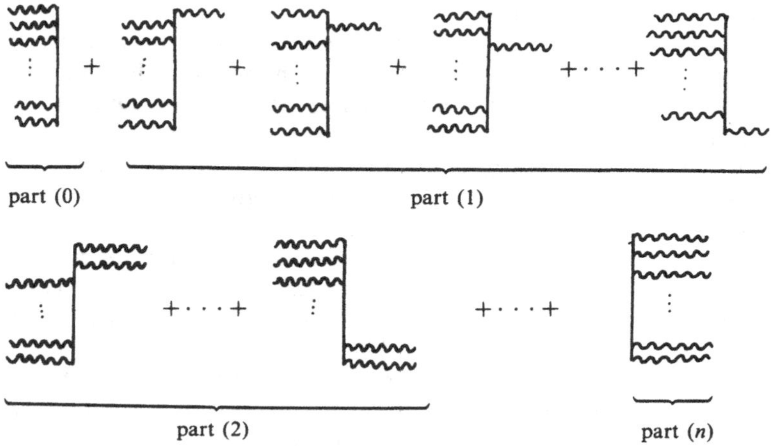

(4.111)

The boson lines to the left of the vertical lines correspond to the creation operators b_r^+, and the boson lines to the right of the vertical lines correspond to annihilation operators b_r. The indices of boson operators in the normal product $\mathcal{N}[\mathbf{B}^n]$ can be arbitrarily permuted, that is,

$$\{\mathcal{N}[(b_{r_1}^+ + b_{r_1})(b_{r_2}^+ + b_{r_2}) \cdots (b_{r_n}^+ + b_{r_n})]\}$$
$$\sim \{\mathcal{N}[(b_{r_{i(1)}}^+ + b_{r_{i(1)}})(b_{r_{i(2)}}^+ + b_{r_{i(2)}}) \cdots (b_{r_{i(n)}}^+ + b_{r_{i(n)}})]\} \quad (4.112)$$

where $i(k)$, $k = 1, \ldots, n$ are the arbitrary permutations of n numbers. Therefore, the individual parts in (4.111) contribute equivalently to the weight factor, and for every part of (4.111) we can write only one skeleton which represents $\binom{n}{k}$ equivalent members in part (k), $k = 1, 2, \ldots, n$. From the individual skeleton we construct a diagram by connecting it with another skeleton in all possible ways. We mean the following. We choose the arbitrary skeleton. If we connect this skeleton with another skeleton to get the diagram, then this chosen skeleton will appear in the following variants:

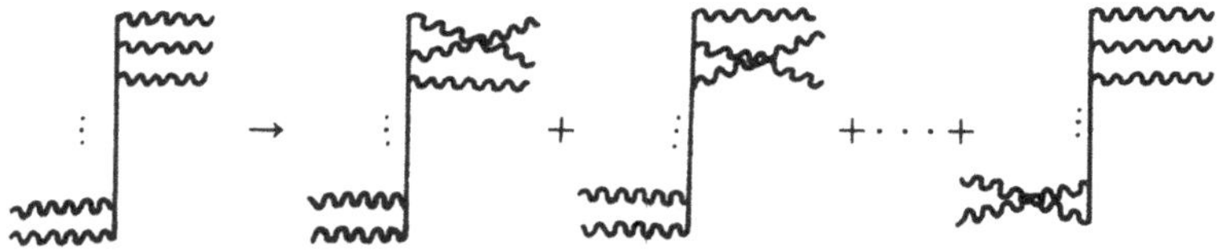

(4.113)

All these variants of one skeleton contribute to the diagram in an equivalent way. For every skeleton corresponding to (k) part (4.111) we have $(n-k)!k!$ variants, and therefore every part (k) is represented in the diagram by $(n-k)!k!\binom{n}{k} = n!$ equivalent contributions, that is, independent on the index k of individual parts.

Let us now imagine a complete vacuum boson diagram in the mth order of the perturbation expansion in which we have m interactions. From every interaction boson vertex outgoing n_i boson lines, $i = 1, \ldots, m$. Due to the permutation of the boson lines the ith interaction vertex contributes to the diagram according to (4.113) by $n_i!$ equivalent contributions. Therefore, we draw coordinate boson interaction vertices as the polygon from which there are outgoing boson lines independently on their order. We only introduce the weight factor $n_i!$ to every interaction boson vertex. This holds, of course, in the special case when two arbitrary boson interaction vertices are connected by one line. In the general case, when the interaction vertex (j) and interaction vertex $(k)(j, k = 1, \ldots, m; j \neq k)$ are connected by n_{jk} lines, we also have to consider the factor $n_{jk}!$ for every pair of vertices (j, k). This is because the boson line shared by two vertices is sufficient to permute only on one of the vertices. Therefore, the weight factor

$$W_B(I) = \prod_{i=1}^{m} n_i! \left(\prod_{\substack{j,k=1 \\ j>k}}^{m} n_{jk}! \right)^{-1} \tag{4.114}$$

corresponds to the boson weight factor of perturbation theory expressions calculated on the basis of the Wick theorem applied to the normal product $N[\mathbf{B}^n]$. We work with the boson operator $\mathbf{B}^{(n)}$ defined by (2.35) as

$$B_{r_1 \cdots r_n} \equiv \frac{1}{n!} N[(b_{r_1}^+ + b_{r_1})(b_{r_2}^+ + b_{r_2}) \cdots (b_{r_n}^+ + b_{r_n})] \tag{4.115}$$

Therefore, we introduce another weight factor

$$W_B(II) = \left(\prod_{i=1}^{m} n_i! \right)^{-1} \tag{4.116}$$

The final weight factor for the boson vacuum diagrams has the form

$$W_B(0) = W_B(I)\, W_B(II) = \left(\prod_{\substack{j,k=1 \\ j>k}}^{m} n_{jk}! \right)^{-1} \tag{4.117}$$

We have the very simple rule that the weight factor depends on the number n_{jk} of the shared pair of lines. As an example we present the weight factor for these three boson vacuum diagrams:

$$W_B(0) = 1 \qquad W_B(0) = \tfrac{1}{2} \qquad W_B(0) = \tfrac{1}{6} \tag{4.118}$$

Finally, we generalize the derivation of the weight factor W_B to more general types of boson diagrams, namely, when the reference bra and ket vectors are not vacuum states but contain the arbitrary number of bosons. We have in mind the diagrams with open boson lines of the form

(4.119)

Let the bra vector contain $n_r^<$ bosons and the ket $n_r^>$ bosons for every vibrational mode r:

$$\langle \text{bra}| = \prod_r \langle n_r^<| = \prod_r \frac{1}{(n_r^<!)^{1/2}} \langle 0 (b_r)^{n_r<} \tag{4.120}$$

$$|\text{ket}\rangle = \prod_r |n_r^>\rangle = \prod_r \frac{1}{(n_r^>!)^{1/2}} (b_{r^+})^{n_r>} |0\rangle \tag{4.121}$$

If we introduce the weight factor W_B^{III} as

$$W_B(III) = \left(\prod_r n_r^<! n_r^>! \right)^{-1/2} \tag{4.122}$$

then we have for the weight factor W_B

$$W_B = W_B(0)\, W_B(III) = \frac{1}{\prod_{\substack{j,k=1 \\ j\neq k}}^{m} n_{jk}! (\prod_r n_r^<! n_r^>!)^{1/2}} \tag{4.123}$$

To assign the matrix element to the particular diagram of the type in (4.108) expected for the described weight we follow the rules for the fermion-type diagrams.[4,7,9,11]

5. Connection of the Technique of Canonical Transformations with the Analytical Derivative Methods

As we said in Section 1, if we want to use the MBPT from (1.22) with the Hamiltonian $H_N(\chi)$ in (1.7), it is necessary to know the creation ($a_P^+(\chi)$) and annihilation ($a_Q(\chi)$) operators. As we have shown in previous sections these operators are for the case of quantization of the V–E Hamiltonian, fully determined by the knowledge of the C_{PQ} (3.5) coefficients. Therefore, there is the question of what is the connection of our canonical transformation method and the inner structure of gradient and/or analytic methods. The C_{PQ} coefficients define the new set of creation (annihilation) operators which create (annihilate) the fermion on the R-dependent spin orbital $|P(R)\rangle$ (3.1):

$$|P(R)\rangle = \sum_Q C_{QP}(R)|Q\rangle \tag{5.1}$$

Note that the $|Q\rangle$ orbitals are R-independent, and the whole dynamics of changing the molecular orbitals $|P(R)\rangle$ is included in C_{PQ} coefficients. The variable R means internuclear distance, and specifically $R = 0$ means the equilibrium distance.

Therefore, we focus our attention on the coefficients C_{PQ} of our canonical transformation. According to (5.1) we have

$$C_{PQ} = C_{PQ}(R) = \langle P(0)|Q(R)\rangle \tag{5.2}$$

Let us write the *MOLCAO* expansions as

$$|P(0)\rangle = \sum_\mu \gamma_{P\mu}(0)|\mu(0)\rangle \tag{5.3}$$

and

$$|P(R)\rangle = \sum_\mu \gamma_{P\mu}(R)|\mu(R)\rangle \tag{5.4}$$

where $\gamma_{P\mu}$ are coefficients of *MOLCAO* expansions, and $|\mu\rangle$ is the basis set of atomic orbitals. In equation (5.4) the change of molecular orbitals reflects the changes of *LCAO* coefficients as well as the change of the atomic orbitals basis set. The motion of the basis sets in (5.1) and (5.4) is identical to the adiabatic motion of the nuclei. As we have said, the crucial step in analytical derivative methods is the calculation of coupled–perturbed Hartree–Fock equations.[14–16]

In *CPHF* equations we proceed as follows. We follow Pople's notation.[16] We generally write the dependence of Hartree–Fock equations on some parameter y as

$$F(y)C_P(y) = E_P(y)S(y)C_P(y) \tag{5.5}$$

where F is the Hartree–Fock operator. Or in matrix form

$$\mathbf{F}(y)\mathbf{C}(y) = \mathbf{S}(y)\mathbf{C}(y)\mathbf{E}(y) \tag{5.6}$$

The orthonormality condition is given as

$$\mathbf{C}^{+}(y)\mathbf{S}(y)\mathbf{C}(y) = \mathbf{1} \tag{5.7}$$

Generally, we have two different *MO* basis sets. The basis set that is y-independent ($y = 0$) we write as an *MOLCAO* expansion (5.3):

$$\chi_P(0) = \sum_{\mu} C_{\mu P}(0)\omega_{\mu}(0) \tag{5.8}$$

We want to know the basis set that is y-dependent, which may be written as

$$\chi_P(y) = \sum_{\mu} C_{\mu P}(y)\omega_{\mu}(y) \tag{5.9}$$

The solution of the *CPHF* equation is achieved through the definition of the new basis set

$$\hat{\chi}_q(y) = \sum_{\mu} C_{\mu q}(0)\omega_{\mu}(y) \tag{5.10}$$

To obtain $\chi_P(y)$ in (5.9) we define

$$\chi_P(y) = \sum_{Q} u_{QP}(y)\hat{\chi}_Q(y) \tag{5.11}$$

Therefore, for the MO expansion coefficient $C(y)$ we write

$$C(y) = C(0)\mathscr{U}(y) \tag{5.12}$$

Substituting (5.12) into (5.6) and (5.7) we can introduce

$$\mathscr{F}(y) = C^{+}(0)F(y)C(0) \tag{5.13}$$

and

$$\mathscr{S}(y) = C^{+}(0)S(y)C(0) \tag{5.14}$$

and rewrite equations (5.6) and (5.7) as

$$\mathscr{F}(y)\mathscr{U}(y) = \mathscr{S}(y)\mathscr{U}(y)E(y) \tag{5.15}$$

and

$$\mathscr{U}^{+}(y)\mathscr{S}(y)\mathscr{U}(y) = 1 \tag{5.16}$$

Now we perform the Taylor expressions for

$$\mathscr{F}(y) = E(0) + y\mathscr{F}^{(1)} + o(y^2) \tag{5.17}$$

$$\mathscr{S}(y) = 1 + y\mathscr{S}^{(1)} + o(y^2) \tag{5.18}$$

$$\mathscr{U}(y) = 1 + y\mathscr{U}^{(1)} + o(y^2) \tag{5.19}$$

$$E(y) = E(0) + yE^{(1)} + o(y^2) \tag{5.20}$$

To bring $\mathscr{F}(y)$ in (5.17) to diagonal form we diagonalize $\mathscr{F}^{(1)}$, etc. To follow our notation we introduce in accordance with (5.10) the new basis set

$$|\hat{P}\rangle = |\hat{P}(R)\rangle = \sum_{\mu} \gamma_{P\mu}(0)|\mu(R)\rangle \tag{5.21}$$

and

$$|P(R)\rangle = \sum_{Q} \hat{C}_{QP}(R)|\hat{Q}\rangle \tag{5.22}$$

Note that $\hat{C}_{QP}(R)$ coefficients are identical to the $u_{QP}(y)$ coefficients (5.11) in Ref. 16. The comparison of our method based on canonical transformation with the analytic derivative method is based on the relation between our C_{PQ} coefficients and $\hat{C}_{PQ}$ coefficients from *CPHF* equations. Moreover, finding this relation will permit us to calculate C_{PQ} coefficients from the solution of *CPHF* equations. Comparing (5.1) with (5.22) we get

$$C_{PQ} = \sum_R \hat{C}_{RQ} \langle P(0)|\hat{R}\rangle \tag{5.23}$$

Because in the approximate Hamiltonian of (4.107) we need to know the C_{PQ} coefficients up to the second order of the Taylor expansion (included), we write

$$C_{PQ}^{(1)} = \hat{C}_{PQ}^{(1)} + \langle P(0)|\hat{Q}\rangle^{(1)} \tag{5.24}$$

$$C_{PQ}^{(2)} = \hat{C}_{PQ}^{(2)} + \sum_R \langle P(0)|\hat{R}\rangle^{(1)} \hat{C}_{RQ}^{(1)} + \langle P(0)|\hat{Q}\rangle^{(2)} \tag{5.25}$$

The upper index means the order of the Taylor expansion and therefore also the corresponding partial derivatives either with respect to Cartesian or normal coordinates. In equations (5.24) and (5.25) we have

$$\langle P(0)|\hat{Q}\rangle^{(k)} = \sum_{\mu\nu} \gamma_{P\mu}(0)\gamma_{Q\nu}(0)\langle \mu(0)|\nu(R)\rangle^{(k)} \tag{5.26}$$

After introducing the relation between coefficients C_{PQ} of the canonical transformation and coefficients $\hat{C}_{PQ}$ from the *CPHF* equations we write the expressions for individual terms of the adiabatic Hamiltonian H_A (4.15) through the coefficients $\hat{C}_{PQ}$:

$$E_{SCF} = \sum_{RSI} \hat{h}_{RS}\hat{C}_{RI}\hat{C}_{SI} + \tfrac{1}{2} \sum_{RSTUIJ} (\hat{v}_{RTSU} - \hat{v}_{RSTU})\hat{C}_{RI}\hat{C}_{SI}\hat{C}_{TJ}\hat{C}_{UJ} \tag{5.27}$$

$$f_{PQ} = \sum_{RS} \hat{h}_{RS}\hat{C}_{RP}\hat{C}_{SQ} + \sum_{RSTUI} (\hat{v}_{RTSU} - \hat{v}_{RSTU})\hat{C}_{RP}\hat{C}_{SQ}\hat{C}_{TI}\hat{C}_{UI} \tag{5.28}$$

$$v_{PQRS} = \sum_{TUVW} \hat{v}_{TUVW}\hat{C}_{TP}\hat{C}_{UQ}\hat{C}_{VR}\hat{C}_{WS} \tag{5.29}$$

where we have used

$$\hat{h}_{PQ} = \langle \hat{P}|h|\hat{Q}\rangle \tag{5.30}$$

$$\hat{v}_{PQRS} = \langle \hat{P}\hat{Q}|v^0|\hat{R}\hat{S}\rangle \tag{5.31}$$

These quantities are given through the individual orders of the Taylor expansion as

$$\hat{h}^{(k)}_{PQ} = \sum_{\mu\nu} \gamma_{P\mu}(0)\gamma_{Q\nu}(0)\langle\mu(R)|h(R)|\nu(R)\rangle^{(k)} \tag{5.32}$$

$$\hat{v}^{(k)}_{PQRS} = \sum_{\mu\nu\rho\tau} \gamma_{P\mu}(0)\gamma_{Q\nu}(0)\gamma_{R\rho}(0)\gamma_{S\tau}(0)\langle\mu(R)\nu(R)|v^0|\rho(R)\tau(R)\rangle^{(k)} \tag{5.33}$$

To simplify what follows, we introduce

$$\hat{E}_{SCF} = \sum_I \hat{h}_{II} + \tfrac{1}{2}\sum_{IJ}(\hat{v}_{IJIJ} - \hat{v}_{IIJJ}) \tag{5.34}$$

$$\hat{f}_{PQ} = \hat{h}_{PQ} + \sum_I (\hat{v}_{PIQI} - \hat{v}_{PQII}) \tag{5.35}$$

$$\hat{g}_{PQ} = \sum_{RI} (2v^o_{PRQI} - v^o_{PQRI} - v^o_{PQIR})\hat{C}_{RI} \tag{5.36}$$

$$\hat{y}_{PQ} = \sum_{RI} (2\hat{v}'_{PRQI} - \hat{v}'_{PQRI} - \hat{v}'_{PQIR})\hat{C}_{RI} \tag{5.37}$$

$$\hat{z}_{PQ} = \sum_{RSI} (v^o_{RPSQ} - v^o_{RSPQ})\hat{C}'_{RI}\hat{C}'_{SI} \tag{5.38}$$

where

$$\hat{v}'_{PQRS} = \hat{v}_{PQRS} - \hat{v}^o_{PQRS} \tag{5.39}$$

In analogy to (3.39) we introduce

$$\hat{C}'_{PQ} = \hat{C}_{PQ} - \delta_{PQ} \tag{5.40}$$

This permits us to obtain the quantities (4.67)–(4.71) from the solution of the *CPHF* equations:

$$f^{(1)}_{PQ} = \hat{f}^{(1)}_{PQ} + \hat{g}^{(1)}_{PQ} + (\epsilon_P - \epsilon_Q)\hat{C}^{(1)}_{PQ} - \epsilon_Q\hat{S}^{(1)}_{PQ} = \epsilon^{(1)}{}_P\delta_{PQ} \tag{5.41}$$

$$\begin{aligned} f^{(2)}_{PQ} &= \hat{f}^{(2)}_{PQ} + \hat{g}^{(2)}_{PQ} + \hat{y}^{(2)}_{PQ} + \hat{z}^{(2)}_{PQ} + \epsilon^{(1)}_P\hat{C}^{(1)}_{PQ} + \epsilon^{(1)}_Q\hat{C}^{(1)}_{PQ} \\ &\quad + \sum_R (\epsilon_P - \epsilon_R)\hat{C}^{(1)}_{RP}\hat{C}^{(1)}_{RQ} + (\epsilon_P - \epsilon_Q)\sum_R \hat{S}^{(1)}_{PR}\hat{C}^{(1)}_{RQ} \\ &\quad - \epsilon_Q\hat{S}^{(2)}_{PQ} + (\epsilon_P - \epsilon_Q)C^{(2)}_{PQ} \\ &= \epsilon^{(2)}_P\,\delta_{PQ} \end{aligned} \tag{5.42}$$

$$v^{(1)}_{PQRS} = \hat{v}^{(1)}_{PQRS} + \sum_T (v^0_{TQRS}\hat{C}^{(1)}_{TP} + v^0_{PTRS}\hat{C}^{(1)}_{TQ} + v^0_{PQTS}\hat{C}^{(1)}_{TR} + v^0_{PQRT}\hat{C}^{(1)}_{TS}) \tag{5.43}$$

$$\begin{aligned} v^{(2)}_{PQRS} = \hat{v}^{(2)}_{PQRS} &+ \sum_T (\hat{v}^{(1)}_{TQRS}\hat{C}^{(1)}_{TP} + \hat{v}^{(1)}_{PTRS}\hat{C}^{(1)}_{TQ} + \hat{v}^{(1)}_{PQTS}\hat{C}^{(1)}_{TR} \\ &+ \hat{v}^{1}_{PQRT}\hat{C}^{(1)}_{TS} + v^0_{TQRS}\hat{C}^{(2)}_{TP} + v^0_{PTRS}\hat{C}^{(2)}_{TQ} + v^0_{PQTS}\hat{C}^{(2)}_{TR} + v^0_{PQRT}\hat{C}^{(2)}_{TS}) \\ &+ \sum_{TU} (v^0_{TURS}\hat{C}^{(1)}_{TP}\hat{C}^{(1)}_{UQ} + v^0_{TQUS}\hat{C}^{(1)}_{TP}\hat{C}^{(1)}_{UR} + v^0_{TQRU}\hat{C}^{(1)}_{TP}\hat{C}^{(1)}_{US} \\ &+ v^0_{PTUS}\hat{C}^{(1)}_{TQ}\hat{C}^{(1)}_{UR} + v^0_{PTRU}\hat{C}^{(1)}_{TQ}\hat{C}^{(1)}_{US} + v^0_{PQTU}\hat{C}^{(1)}_{TR}\hat{C}^{(1)}_{US}) \end{aligned} \tag{5.44}$$

$$E^{(1)}_{SCF} = \hat{E}^{(1)}_{SCF} - \sum_I \epsilon_I \hat{S}^{(1)}_{II} \tag{5.45}$$

$$E^{(2)}_{SCF} = \hat{E}^{(2)}_{SCF} - \sum_I (\epsilon_I \hat{S}^{(2)}_{II} + \tfrac{1}{2}\epsilon^{(1)}_I \hat{S}^{(1)}_{II}) + \sum_{RI} (\hat{f}^{(1)}_{RI} - \epsilon_I \hat{S}^{(1)}_{RI})\hat{C}^{(1)}_{RI} \tag{5.46}$$

$$\begin{aligned} E^{(3)}_{SCF} = \hat{E}^{(3)}_{SCF} &- \sum_I (\epsilon_I \hat{S}^{(3)}_{II} + \tfrac{1}{2}\epsilon^{(1)}_I \hat{S}^{(2)}_{II} + \tfrac{1}{2}\epsilon^{(2)}_I \hat{S}^{(1)}_{II}) \\ &+ \sum_{RI} [(\hat{f}^{(2)}_{RI} - \hat{z}^{(2)}_{RI} - \epsilon_I \hat{S}^{(2)}_{RI} + \tfrac{1}{2}\epsilon^{(1)}_I \hat{C}^{(1)}_{RI})\hat{C}^{(1)}_{RI} + (\hat{f}^{(1)}_{RI} - \epsilon_I \hat{S}^{(1)}_{RI})\hat{C}^{(2)}_{RI}] \end{aligned} \tag{5.47}$$

To conclude this section we use the properties of (3.45) and (3.46). That is, we express the d_{rPQ} coefficients of the boson canonical transformation in (3.15) through the C_{PQ} coefficients, namely,

$$d^{(0)}_{rPQ} = C^r_{PQ} \tag{5.48}$$

and

$$d^s_{rPQ} = C^{rs}_{PQ} + \sum_R C^{s*}_{RP} C^r_{RQ} \tag{5.49}$$

6. Calculation of the Energies of the First Vibrational Transitions in Molecular Systems

6.1. Calculation of Energies

In this section we use the Hamiltonian in (4.107) to derive the explicit expressions for the calculation of the energies of the first vibrational transitions in molecules through the many-body perturbation theory. We also compare our approach with that based on the analytical derivative methods.

The diagrammatic technique permits us to make the theory rather transparent and to distinguish between various perturbation contributions. The energy of the first vibrational transition ε_1 is defined as an energy difference $E_1 - E_0$ where E_1 is the total energy with an excited phonon (boson) and E_0 is the total energy without an excited phonon

$$\varepsilon_1 = E_1 - E_0 \tag{6.1}$$

For the calculation of the total energies, E_1 and E_0, we use the non-degenerate Rayleigh–Schrödinger *MBPT* expansions. Let us assume that we know the solution of the unperturbed Schrödinger equation

$$H_0|\varphi_i\rangle = e_i|\varphi_i\rangle \tag{6.2}$$

where H_0 is the unperturbed part of the Hamiltonian (4.107). We want to know the solution of perturbed (exact) Schrödinger equation

$$H|\Phi_i\rangle = E_i|\Phi_i\rangle \tag{6.3}$$

where H is the Hamiltonian from (4.107). The perturbed energy E_i will be given through the *MBPT* expansion as

$$E_i = e_i + \langle\varphi_i|H'|\varphi_i\rangle + \langle\varphi_i|H'Q_iH'|\varphi_i\rangle + \cdots \tag{6.4}$$

where H' is the perturbation in (4.107) and Q_i is the resolvent

$$Q_j = \sum_{i \neq j} \frac{|\varphi_i\rangle\langle\varphi_i|}{e_i - e_j} \tag{6.5}$$

Because our sets of boson creation and annihilation operators and fermion creation and annihilation operators commute, we write our unperturbed wave function $|\varphi_i\rangle$ as the product of the fermion state vector $|\psi_i\rangle$ and the boson state vector $|\chi_i\rangle$, that is,

$$|\varphi_i\rangle = |\psi_i\rangle|\chi_i\rangle \tag{6.6}$$

We shall study only the first vibrational transitions, that is, a one-phonon transition where our fermion reference state $|\psi_i\rangle$ will be the lowest *SCF* state

$|\psi_0\rangle$. The phonon reference vector $|\chi_i\rangle$ is chosen for the lowest state as the boson vacuum $|0\rangle$, that is,

$$|\chi_0(0)\rangle = |0\rangle \tag{6.7}$$

The one-phonon excited state reference vector $|\chi_1(r)\rangle$ is chosen as

$$|\chi_1(r)\rangle = b_r^+|0\rangle \tag{6.8}$$

The initial unperturbed state vector for the system without an excited phonon is chosen as

$$|\varphi_0(0)\rangle = |\psi_0\rangle|\chi_0(0)\rangle \tag{6.9}$$

and the unperturbed state vector for the system with an excited phonon is given as

$$|\varphi_1(r)\rangle = |\psi_0\rangle|\chi_1(r)\rangle \tag{6.10}$$

Adopting the perturbation expansion (6.4) for E_1 as well as for E_0 in equation (6.1), we can write

$$\begin{aligned}\varepsilon_1 = \{&e_1 + \langle\varphi_1(r)|H'|\varphi_1(r)\rangle + \langle\varphi_1(r)|H'Q_1H'|\varphi_1(r)\rangle + \cdots\}\\ &- \{e_0 + \langle\varphi_0(0)|H'|\varphi_0(0)\rangle + \langle\varphi_0(0)|H'Q_0H'|\varphi_0(0)\rangle + \cdots\}\end{aligned} \tag{6.11}$$

Substituting (6.8)–(6.10) into (6.11) we have

$$\begin{aligned}\varepsilon_1 = \{&e_1 + \langle\psi_0|\langle 0|b_r|H'|b_r^+|0\rangle|\psi_0\rangle + \langle\psi_0|\langle 0|b_r|H'Q_1H'|b_r^+|0\rangle|\psi_0\rangle + \cdots\}\\ &- \{e_0 + \langle\psi_0|\langle 0|H'|0\rangle\psi_0\rangle + \langle\psi_0|\langle 0|H'Q_0H'|0\rangle|\psi_0\rangle + \cdots\}\end{aligned} \tag{6.12}$$

Equation (6.12) can be treated diagrammatically. We present the exact cancellation of diagrams which originate from the E_0 term with the part of the diagrams which originate from the E_1 term. Due to this cancellation the remaining diagrams give us the explicit formula for the energy of the first vibrational transition. Therefore, the energy of the first vibrational transition is not given as the subtraction of two numbers but as a single lucid formula. To understand better this cancellation we refer the reader to the papers where the calculations of the energy differences by *MBPT* are presented, for example, calculation of ionization potentials.[7,9,27,28] The cancellation presented here is of the same nature.

Equation (6.12) can be expressed diagrammatically in the following manner

$$\varepsilon_1 = \{e_1 + \text{〰▨〰}\} - \{e_0 + \text{▨}\} \qquad (6.13)$$

where the boson line 〰〰 stands for the $b_r(b_r^+)$ operator, and the box stands for the diagrams that originate from the perturbations. The second term on the right-hand side of equation (6.13) can be treated schematically as follows

$$\{ \text{〰 ▨ 〰} \} = \{ \text{▨} \} + \{ \text{〰▨〰} \} \qquad (6.14)$$

where the first term on the right-hand side corresponds to the contraction of two boson $b_r(b_r^+)$ lines, and the second term represents all possible Feynman-like diagrams where the boson lines $b_r(b_r^+)$ are connected with the perturbation diagrams. Because the contraction of two boson lines $b_r(b_r^+)$ corresponds to the multiplication of box ▨ by number one we have

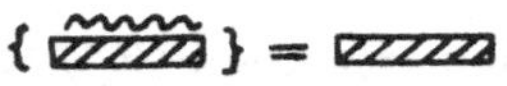

$$\{ \text{▨} \} = \text{▨} \qquad (6.15)$$

Therefore, substituting (6.14) and (6.15) into (6.13) we have

$$\varepsilon_1 = \hbar\omega_r + \text{〰▨〰} \qquad (6.16)$$

where we use the fact that

$$e_1 - e_0 = \hbar\omega_r \qquad (6.17)$$

Equation (6.16) gives us the explicit formula for the energy of the first vibrational transition, and the second term on the right-hand side corresponds to the correction to the harmonic frequencies $\hbar\omega_r$. We construct all possible diagrams from equation (6.16) which originate from the Hamiltonian in (4.107). Because the Hamiltonian in (4.107) contains many perturbation terms we calculate only the low-order $MBPT$ contribution. We label the corrections to harmonic frequencies $\hbar\omega_r$ as $\Delta\varepsilon$. Equation (6.16) then has the form

$$\varepsilon_1 = \hbar\omega_r + \Delta\varepsilon \qquad (6.18)$$

where by $\Delta\varepsilon$ we have in mind the summation of all diagrammatic contributions

$$\Delta\varepsilon = \sum_x \Delta\varepsilon_x \tag{6.19}$$

The following Hugenholtz-type diagrams correspond to the corrections to the harmonic frequencies due to the fermion correlation effects.[24,29]

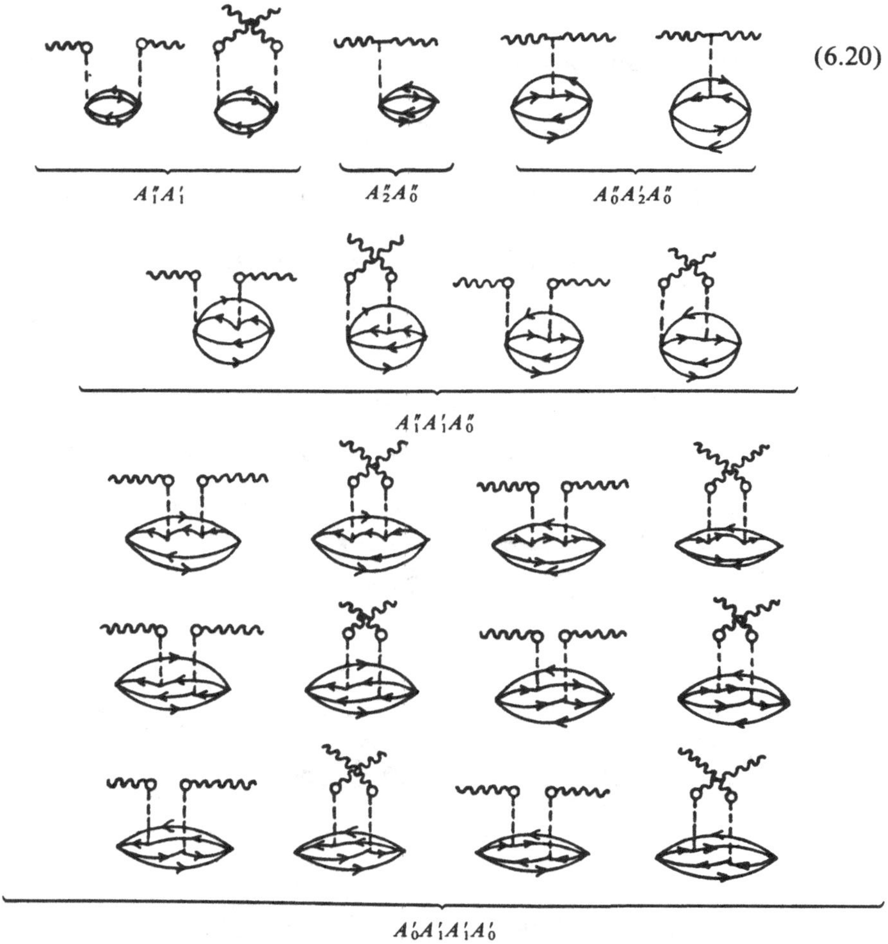

(6.20)

Using the diagrammatic rules described in Section 5 we get the following matrix elements in spin orbitals for the individual diagrams. To simplify the formulae we adopt the following notation for the denominators

$$\Delta^{I_1\cdots I_\alpha r_1\cdots r_\gamma}_{A_1\cdots A_\beta s_1\cdots s_\delta} = \left(\sum_{k=1}^{\alpha}\varepsilon_{I_k} - \sum_{k=1}^{\beta}\varepsilon_{A_k} + \sum_{k=1}^{\gamma}\hbar\omega_{r_k} - \sum_{k=1}^{\delta}\hbar\omega_{s_k}\right)^{-1} \tag{6.21}$$

The number (2) means the multiplicity of the diagram.

$$\Delta e(r)_{A_1''A_1''} = \frac{1}{2}\sum_{ABIJ} v^r_{ABIJ}(\Delta^{IJr}_{AB} + \Delta^{IJ}_{ABr})(v^r_{ABIJ} - v^r_{ABJI}) \tag{6.22}$$

$$\Delta e(r)_{A_2''A_0''} = \sum_{ABIJ}^{rr} v^{rr}_{ABIJ}\Delta^{IJ}_{AB}(v^0_{ABIJ} - v^0_{ABJI}) \tag{6.23}$$

$$\Delta e(r)_{A_1''A_1''A_0''} = 2\sum_{ABIJ} v^r_{ABIJ}(\Delta^{IJr}_{AB} + \Delta^{IJ}_{ABr})(\varepsilon^r_A - \varepsilon^r_I)\Delta^{IJ}_{AB}(v^0_{ABIJ} - v^0_{ABJI}) \tag{6.24}$$

$$\Delta e(r)_{A_0''A_2''A_0''} = \sum_{ABIJ} v^0_{ABIJ}(\Delta^{IJ}_{AB})^2(\varepsilon^{rr}_A - \varepsilon^{rr}_I)(v^0_{ABIJ} - v^0_{ABJI}) \tag{6.25}$$

$$\Delta e(r)_{A_0''A_1''A_1''A_0''} = \sum_{ABIJ} v^0_{ABIJ}(\Delta^{IJ}_{AB})^2(\varepsilon^r_A - \varepsilon^r_I)(\varepsilon^r_A + \varepsilon^r_B - \varepsilon^r_I - \varepsilon^r_J) \cdot (\Delta^{IJr}_{AB} + \Delta^{IJ}_{ABr})(v^0_{ABIJ} - v^0_{ABJI}) \tag{6.26}$$

6.1.1. *Corrections Due to Anharmonicity Effects*

The following diagrams correspond to anharmonicity corrections.

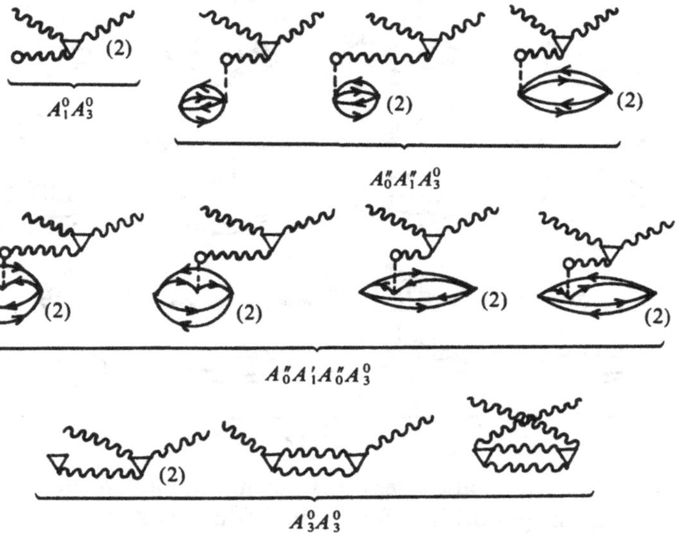

(6.27)

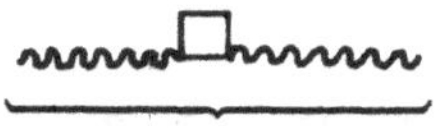

A_4^0

The corresponding matrix elements are

$$\Delta e(r)_{A_1^0 A_3^0} = 2\sum_s E^s \Delta_s E^{rrs} \tag{6.28}$$

$$\Delta e(r)_{A_0'' A_1'' A_3^0} = \sum_{ABIJs} (v^0_{ABIJ} - v^0_{ABJI})[\Delta^{IJ}_{AB}\Delta_s + \Delta^{IJ}_{ABs}(\Delta_s + \Delta^{IJ}_{AB})] v^s_{ABIJ} E^{rrs} \tag{6.29}$$

$$\Delta e(r)_{A_0'' A_1'' A_0'' A_3^0} = 2\sum_{ABIJs} v^0_{ABIJ}\Delta^{IJ}_{AB}(\varepsilon^s_A - \varepsilon^s_I)\Delta^{IJ}_{ABs}(v^0_{ABIJ} - v^0_{ABJI})(\Delta_s + \Delta^{IJ}_{AB})E^{rrs} \tag{6.30}$$

$$\Delta e(r)_{A_3^0 A_3^0} = 2\sum_s E^{00s}\Delta_s E^{rrs} + \tfrac{1}{2}\sum_{ts} (E^{rst})^2(\Delta^r_{st} + \Delta_{rst}) \tag{6.31}$$

$$\Delta e(r)_{A_4^0} = E^{00rr} \tag{6.32}$$

6.1.2. *Corrections Due to Nonadiabatic Effects*

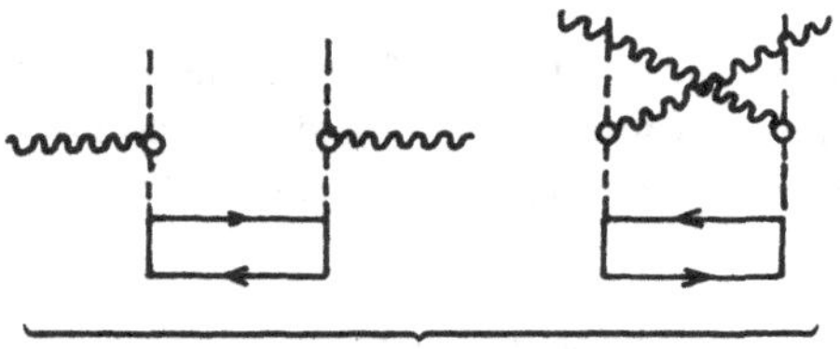

(6.33)

$B_1^* B_1^*$

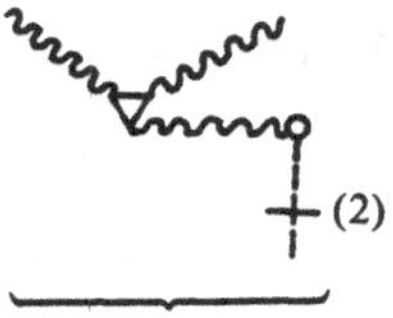

(6.34)

$A_3^0 B_3^0$

The corresponding matrix elements are

$$\Delta e(r)_{B_1^* B_1^*} = \sum_{AI} (m^{0,r}_{AI})^2(\Delta^{Ir}_A + \Delta^I_{Ar}) \tag{6.35}$$

$$\Delta e(r)_{A_3^0 B_3^0} = 2\sum_s E^{rrs}\Delta_s \tilde{E}^{s,00} \tag{6.36}$$

6.2. Computational Details

The present formulae were used to calculate the energies of the first vibrational transitions for some simple molecular systems. The calculated corrections to harmonic frequencies are due to electron correlation and nonadiabatic effects. To calculate the corrections due to anharmonicity effects we need to know the third and the fourth derivatives of the SCF energy. Our aim is to test the presented theory of the canonical transformations. We compare our results with those obtained by the gradient techniques. If we look carefully at the formulae for the corrections due to the correction effects we see that our formulae are identical to the diagonal second derivative of the second order of the $MBPT$.[30] To include the off-diagonal contributions we formulate the theory through the quasi-degenerate many-body perturbation theory[27,31-35] which generalizes the previous approach for the cases when $\omega_r \approx \omega_s$ and/or $\omega_r = \omega_s$. It is convenient to choose the model space as a complete space of one-boson states.[36] We get the secular determinant of the order $N \times N$, where N is the number of the vibrational modes. This means that we solve the determinant

$$\det|\langle r|H_{\text{eff}}|s\rangle - E_1\,\delta_{rs}| = 0 \qquad (6.37)$$

where E_1 is the total energy with one excited phonon, and H_{eff} is the effective Hamiltonian.

There are many versions of the effective Hamiltonians. Their mutual relations as well as the notation used here is described in our recent paper.[35] If we denote for simplicity the matrix elements of the effective Hamiltonian in (6.37) as

$$\langle r|H_{\text{eff}}|s\rangle = E(r - s) \qquad (6.38)$$

and further separate them onto the unperturbed part and the perturbation and formulate the effective Hamiltonian for the energy difference

$$E(r - s) - E(0)\delta_{rs} = [e_0(r) - e_0(0)]\delta_{rs} + \Delta\varepsilon(r - s)$$
$$- \Delta\varepsilon(0)\delta_{rs} = e_0(r)\delta_{rs} + \Delta e(r - s) \qquad (6.39)$$

The expression $\Delta\varepsilon(r - s) - \Delta\varepsilon(0)\delta_{rs}$ can be represented diagrammatically as

$$\left\{\cdots\!\!\!\!\!\!\!\overset{r}{\sim}\,\boxed{/\!/\!/\!/}\,\overset{s}{\sim}\!\!\!\cdots\right\} - \left\{\boxed{/\!/\!/\!/}\right\}\delta_{rs} = \left\{\overset{r=s}{\boxed{/\!/\!/\!/}}\right\}\delta_{rs} + \left\{\cdots\!\!\overset{r}{\sim}\,\boxed{/\!/\!/\!/}\,\overset{s}{\sim}\!\!\cdots\right\} - \qquad (6.40)$$

$$- \left\{\boxed{/\!/\!/\!/}\right\}\delta_{rs} = \left\{\overset{r}{\sim}\!\!\!\cdots\,\boxed{/\!/\!/\!/}\,\cdots\!\!\!\overset{s}{\sim}\right\} = \Delta e\,(r-s)$$

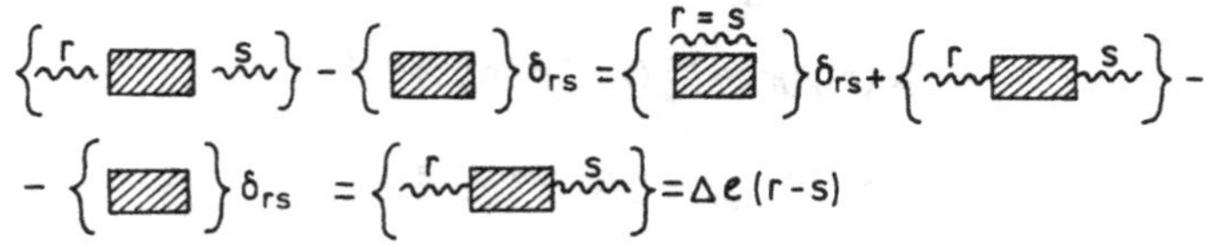

The secular determinant for the energy of the first vibrational transition ε_1 has the form

$$\det|e_0(r)\,\delta_{rs} + \Delta e(r-s) - \varepsilon_1\,\delta_{rs}| = 0 \tag{6.41}$$

The expressions $\Delta e(r-s)$ are expressed diagrammatically as

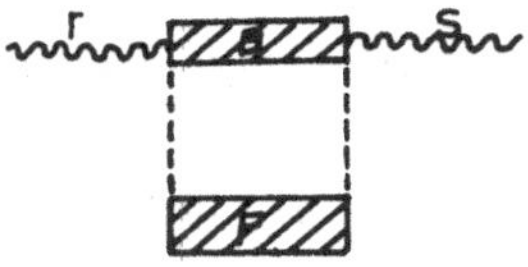

(6.42)

We see that the diagrams are the same as in the case of equation (6.16) except that the external boson lines are not diagonal $(r-r)$ but generally are denoted as $(r-s)$. The explicit expressions derived from these diagrams are the following:

$$\Delta e(r-s)_{A_1''A_1''} = \tfrac{1}{4}\sum_{ABIJ} v^r_{ABIJ}(\Delta^{IJr}_{AB} + \Delta^{IJs}_{AB} + \Delta^{IJ}_{ABr} + \Delta^{IJ}_{ABs})(v^s_{ABIJ} - v^s_{ABJI}) \tag{6.43}$$

$$\Delta e(r-s)_{A_2''A_0''} = \tfrac{1}{4}\sum_{ABIJ} v^{rs}_{ABIJ}(2\Delta^{IJ}_{AB} + \Delta^{IJr}_{ABs} + \Delta^{IJs}_{ABr})(v^0_{ABIJ} - v^0_{ABJI}) \tag{6.44}$$

$$\begin{aligned}\Delta e(r-s)_{A_1''A_1'A_0''} = \tfrac{1}{2}\sum_{ABIJ} \{ & v^r_{ABIJ}[(\Delta^{IJs}_{AB} + \Delta^{IJ}_{ABs})\Delta^{IJ}_{AB} + \Delta^{IJr}_{AB}\Delta^{IJr}_{ABs} \\ & + \Delta^{IJ}_{ABr}\Delta^{IJs}_{ABr}](\varepsilon^s_A - \varepsilon^s_I) + v^s_{ABIJ}[(\Delta^{IJr}_{AB} + \Delta^{IJ}_{ABr})\Delta^{IJ}_{AB} \\ & + \Delta^{IJs}_{AB}\Delta^{IJs}_{ABr} + \Delta^{IJ}_{ABs}\Delta^{IJr}_{ABs}](\varepsilon^r_A - \varepsilon^r_I)\}(v^0_{ABIJ} - v^0_{ABJI})\end{aligned} \tag{6.45}$$

$$\Delta e(r-s)_{A_0''A_2''A_0''} = \tfrac{1}{2}\sum_{ABIJ} v^0_{ABIJ}\Delta^{IJ}_{AB}(\Delta^{IJr}_{ABs} + \Delta^{IJs}_{ABr})(\varepsilon^{rs}_A - \varepsilon^{rs}_I)(v^0_{ABIJ} - v^0_{ABJI}) \tag{6.46}$$

$$\begin{aligned}\Delta e(r-s)_{A_0''A_1''A_1'A_0''} = \tfrac{1}{2}\sum_{ABIJ} & v^0_{ABIJ}\Delta^{IJ}_{AB}[\Delta^{IJr}_{ABs}(\Delta^{IJr}_{AB} + \Delta^{IJ}_{ABs}) + \Delta^{IJs}_{ABr}(\Delta^{IJs}_{AB} + \Delta^{IJ}_{ABr})] \\ & \cdot(\varepsilon^r_A - \varepsilon^r_I)(\varepsilon^s_A + \varepsilon^s_B - \varepsilon^s_I - \varepsilon^s_J)(v^0_{ABIJ} - v^0_{ABJI})\end{aligned} \tag{6.47}$$

$$\Delta e(r-s)_{A_1^0A_3^0} = \tfrac{1}{2}\sum_t E^t(2\Delta_t + \Delta^r_{st} + \Delta^s_{rt})E^{rst} \tag{6.48}$$

$$\begin{aligned}\Delta e(r-s)_{A_0''A_1''A_0''} = \tfrac{1}{4}\sum_{ABIJt} & (v^0_{ABIJ} - v^0_{ABJI})[2(\Delta^{IJ}_{AB} + \Delta^{IJ}_{ABt})\Delta_t \\ & + \Delta^{IJr}_{ABs}(\Delta^r_{st} + \Delta^{IJ}_{ABt}) + \Delta^{IJs}_{ABr}(\Delta^s_{rt} + \Delta^{IJ}_{ABt}) \\ & + \Delta^{IJr}_{ABst}(\Delta^r_{st} + \Delta^{IJ}_{AB}) + \Delta^{IJs}_{ABrt}(\Delta^s_{rt} + \Delta^{IJ}_{AB})]v^t_{ABIJ}E^{rst}\end{aligned} \tag{6.49}$$

$$\Delta e(r-s)_{A_0''A_1''A_0''A_3^0} = \frac{1}{2} \sum_{ABIJt} v^0_{ABIJ}[\Delta^{IJ}_{AB}\Delta^{IJ}_{ABt}(2\Delta_t + \Delta^{IJs}_{ABr} + \Delta^{IJr}_{ABs})$$

$$+ \Delta^{IJr}_{ABs}\Delta^{IJr}_{ABst}(\Delta^r_{st} + \Delta^{IJ}_{AB}) + \Delta^{IJs}_{ABr}\Delta^{IJs}_{ABrt}(\Delta^s_{rt} + \Delta^{IJ}_{AB})]$$

$$\cdot (\varepsilon^t_A - \varepsilon^t_I)(v^0_{ABIJ} - v^0_{ABJI})E^{rst} \tag{6.50}$$

$$\Delta e(r-s)_{A_3^0A_3^0} = \frac{1}{2} \sum_t E^{00t}(2\Delta_t + \Delta^r_{st} + \Delta^s_{rt})E^{rst}$$

$$+ \frac{1}{4} \sum_{tu} E^{rtu}(\Delta^r_{tu} + \Delta^s_{tu} + \Delta_{rtu} + \Delta_{stu})E^{stu} \tag{6.51}$$

$$\Delta e(r-s)_{A_4^0} = E^{00rs} \tag{6.52}$$

$$\Delta e(r-s)_{B_1^\bullet B_1^\bullet} = \frac{1}{2} \sum_{AI} m^{0,r}_{AI}(\Delta^{Ir}_A + \Delta^I_{Ar} + \Delta^{Is}_A + \Delta^I_{As})m^{0,s}_{AI} \tag{6.53}$$

$$\Delta e(r-s)_{A_3^0B_3^0} = \frac{1}{2} \sum_t E^{rst}(2\Delta_t + \Delta^r_{st} + \Delta^s_{rt})\tilde{E}^{t,00} \tag{6.54}$$

In our calculations we present the corrections to the harmonic frequencies that are due to formula (6.16). As we said at the very beginning, we want to compare our theory with the theory based on the gradient techniques. Therefore, we present here the results calculated with the diagonal $MBPT(2)$ force constants and, for completeness, also with the full $MBPT(2)$ force constants. The corrections due to the nonadiabatic effects are calculated only through formula (6.35). To make this formula more transparent we can substitute into (6.35) for

$$m^{0,r}_{AI} = \hbar\omega_r d^0_{rAI} \tag{6.55}$$

and for

$$d^0_{rAI} = \hat{C}^r_{AI} + \langle A(0)|\hat{I}^r\rangle \tag{6.56}$$

and we get the very simple relation

$$\Delta e(r)_{B_1^\bullet B_1^\bullet} = \sum_{AI} \hbar^2\omega_r^2(\hat{C}^r_{AI} + \langle A(0)|\hat{I}^r\rangle)^2(\Delta^{Ir}_A + \Delta^I_{Ar}) \tag{6.57}$$

The molecules calculated are H_2, HD, D_2, H_2O, and H_2CO. All molecules are calculated in a DZP basis set. Dunning's contracted double-zeta bases were used for hydrogen, nitrogen, oxygen, and carbon[37] (a 1.2 scale factor is used in the hydrogen basis) augmented by the optimum polarization functions (hydrogen p-functions, exponent = 0.7; nitrogen d-function, exponent = 0.902; oxygen d-function, exponent = 1.211; carbon d-function,

Table 1. Corrections to *SCF* Harmonic Frequencies Due to Electron Correlation and Nonadiabatic Effects for the H_2O, HD and D_2 Molecules[a,b]

A	*B*	*C*	*D*	*E*	*F*	*G*
H_2	4636.7[c]	−1.2139	4635	4635	−0.496430	
	4585.0[d]	−26.7738	4558	4558	−0.642409	−0.8363
HD	4016.0[c]				−0.286797	
	3971.3[d]				−0.371108	−0.5508
D_2	3280.0[c]				−0.175714	
	3243.4[d]				−0.227356	−0.3048

KEY: *A*, molecule; *B*, *SCF* harmonic frequencies; *C*, corrections due to correlation effects—formulae (6.22–6.26); *D*, harmonic frequencies with the inclusion of the electron correlation effects (this method); *E*, harmonic frequencies calculated by gradient techniques with diagonal *MBPT*(2) force constants; *F*, nonadiabatic correlations due to formula (6.57); *G*, nonadiabatic corrections from Refs. 40 and 41.
[a] All entries are in cm^{-1}.
[b] Geometrics are optimized at the *SCF* level
[c] DZP basis set.
[d] Extended [6*s* 2*p* 1*d*] basis set.[39]

exponent = 0.654) of Redmon et al.[38] The H_2, HD, and D_2 molecules are also calculated with Davidson's basis sets. The extended hydrogen basis set used was [6*s* 2*p* 1*d*].[39]

6.3. Results

Here we only summarize the main results which are published in detail in our recent paper.[29] The results are presented in Tables 1 and 2.

Table 2. Corrections to *SCF* Harmonic Frequencies Due to Electron Correlation and Nonadiabatic Effects for the H_2O and H_2CO Molecules[a,b]

A		*B*	*C*	*D*	*E*	*F*
H_2O	B_2	4258.54	+4.6054	4264	4289	−0.307840
	A_1	1750.51	−37.9544	1712	1590	−0.088145
	A_1	4147.15	−7.0303	4140	4141	−0.284387
H_2CO	A_1	2006.1	−28.7987	1977	1971	−0.153156
	B_1	1330.9	−74.8721	1256	1185	−0.053292
	B_2	3206.0	−18.5513	3187	3180	−0.173960
	A_1	1651.9	−17.5497	1634	1585	−0.123592
	A_1	3130.3	−15.8248	3114	3107	−0.272782
	B_2	1367.2	−55.6957	1312	1284	−0.59555

KEY: *A*, molecule and the irreducible representations of the corresponding symmetry group; *B*, *SCF* harmonic frequencies; *C*, corrections due to correlation effects—formulae (6.22–6.26); *D*, harmonic frequencies with the inclusion of electron correlation (this method); *E*, harmonic frequencies calculated by gradient techniques with complete *MBPT*(2) force constants; *F*, nonadiabatic corrections due to formula (6.57).
[a] All entries are in cm^{-1}.
[b] Geometries are optimized at the *SCF* level.

6.4. Discussion

As mentioned in the introduction the aim of the present chapter has been to demonstrate the technique of canonical transformations in quantization of the V–E Hamiltonian. We have compared the techniques of canonical transformations with the analytical derivative techniques. Moreover, using the technique of Feynman-like diagrams, the many-body perturbation theory based on such developed quantization of the V–E Hamiltonian is rather transparent. We were able to split the V–E Hamiltonian into an unperturbed part and the perturbation so that the many-body Rayleigh–Schrödinger perturbation theory can be fully adopted. The perturbations are due to the electron correlation, anharmonicity, and the nonadiabaticity effects. Using the MBPT approach we derived the explicit expressions for the calculations of the energies of the first vibrational transitions in molecules. The theory is constructed in a way that the harmonic frequencies are calculated on the *SCF* level, and the perturbation corrections are due to electron correlation, anharmonicity effects, and the nonadiabaticity effects.

Because of the approximation adopted for the V–E Hamiltonian in (4.107) the formulae for the correction to the *SCF* harmonic frequencies due to electron correlation (6.22)–(6.26) are identical with the diagonal second derivative of the second order of the many-body perturbation theory MBPT(2). To compare our method with results obtained using the analytical second-order MBPT second-derivative method,[30] we replaced the *SCF* force constants with the diagonal MBPT(2) force constants as well as with the complete MBPT(2) force constants. As seen from Tables 1 and 2 the agreement between our method and that based on the analytical derivative method is very good for all molecules calculated. The nonadiabatic corrections to the *SCF* harmonic frequencies due to formula (6.57) are calculated for all molecules. These are compared with the exact values from Refs. 40 and 41 for H_2, HD, and D_2 molecules (Table 1). As we can see from Table 1 the results are strongly basis set dependent, but for the extended basis set they agree well with the results obtained in Refs. 40 and 41, representing approximately 70% of the exact values. To get better agreement we would have to calculate the contribution from the second term (6.36) as well as higher orders of the perturbation expansion. For the simple approximation adopted we think that the agreement is quite good. Note that to calculate the contribution according to formula (6.57) it is necessary to know only the solution of the first-order CPHF equations.

7. Nonadiabatic Representation of the V–E Hamiltonian

For a great deal of quantum-chemical calculation the adiabatic approximation is sufficient. The nuclear motion takes place on an effective potential surface that is the sum of the electronic energy and the nuclear repulsion as a function of the nuclear coordinates. As is well known, the Born–Oppenheimer approximation[42] is based on the fact that the spacing of electronic eigenvalues is generally large compared with the typical spacing associated with nuclear motion.[43] In other words, this means that the vibrational quantum $\hbar\omega$ is much smaller than the energy difference $E_k - E_l$ between two electronic states. This is the case of nondegeneracy of electronic states. When this condition is violated, the residual coupling via the nuclear kinetic energy operator causes transitions between the adiabatic electronic states. Usually only two or at most a few electronic states are close in energy. In these cases it is necessary to use methods that abandon the Born–Oppenheimer approximation.

Köppel et al.[43] discuss the concept of how to handle non-Born–Oppenheimer phenomena through the introduction of suitable "diabatic" states that may cross as a function of the internuclear distance, whereas the adiabatic electronic states are subjected to the non-crossing rule. In Section 6 we developed the method for the calculation of the energies of the first vibrational transitions in molecules in the adiabatic representation using the many-body perturbation theory. The corrections to the harmonic frequencies that are calculated on the *SCF* level are due to the electron correlation, anharmonicity, and nonadiabatic effects. The nonadiabatic corrections are small for most molecules. In the case of quasi-degeneracy or degeneracy of electronic levels we cannot calculate the nonadiabatic corrections using the perturbation theory. This is the case when potential energy surfaces come close together. The vibrational energies in such cases are usually calculated using the concept of an effective Hamiltonian and diabatic wave functions.[43]

In this section we try to extend the technique of canonical transformations for the case when the Born–Oppenheimer approximation fails. The basic idea is the same as in the case of adiabatic representation where we started from the crude representation in which the V–E Hamiltonian was expressed through the standard electron and phonon operators. Then using the canonical transformations we arrived at the V–E Hamiltonian in which the new fermions correspond to electrons that adiabatically follow the motion of nuclei. The new boson operators correspond to renormalized phonons where the renormalization brings the nonadiabatic corrections.

The transformation from the old set of electron and phonon operators to the new set of fermion and boson operators was described in Section 4.1 and was given as

$$\bar{a}_P = \sum_Q C_{PQ}(B) a_Q \tag{7.1}$$

$$\bar{a}_P^+ = \sum_Q C_{PQ}(B)^+ a_Q^+ \tag{7.2}$$

$$\bar{b} = b_r + \sum_{PQ} d_{rPQ}(B) a_P^+ a_Q \tag{7.3}$$

$$\bar{b}_r^+ = b_r^+ + \sum_{PQ} d_{rQP}(B)^+ a_P^+ a_Q \tag{7.4}$$

In the case of nonadiabatic representation we try to generalize the transformations (7.1)–(7.4). The generalization can be done in a way that from operator coefficients $C_{PQ}(B)$ we arrive at general coefficients $C_{PQ}(b, b^+)$. A similar idea was studied by Wagner.[44] Let us introduce in analogy with the normal coordinate operator B_r (2.30) the momentum operator $\tilde{B}_r$ as

$$\tilde{B}_r = b_r - b_r^+ \tag{7.5}$$

Then we can generalize the equation (7.1) in the following way[36]

$$\bar{a}_P = \sum_Q C_{PQ}(B, \tilde{B}) a_Q \tag{7.6}$$

The new fermions represent the electrons that move in dependence on coordinate and on momentum of nuclei. The electrons due to their finite mass do not follow adiabatically the motion of nuclei, but their motion is phase-shifted with respect to the motion of nuclei.

The general transformation in (7.6) is complex and brings rather complex conditions for the general coefficients $C_{PQ}(B, \tilde{B})$. Instead of using the general transformation in (7.6) we try its simple version in which the general coefficients $C_{PQ}(B, \tilde{B})$ are expressed as a simple product of $C_{PQ}(B)$ and $\tilde{C}_{PQ}(\tilde{B})$ coefficients.[36] The coefficients $C_{PQ}(B)$ correspond to adiabatic transformation as described in Section 6, and the coefficients $\tilde{C}_{PQ}(\tilde{B})$ correspond to the new transformation which we call nonadiabatic. The final nonadiabatic representation will be given by the combination of both transformations. The nonadiabatic transformation can be formulated analogously as the adiabatic transformation in (4.5)–(4.8) except that the coefficients $C_{PQ}(B)$ and $d_{rPQ}(B)$ are substituted for $C_{PQ}(\tilde{B})$ coefficients which are functions of momentum operators $\tilde{B}_r$. The transformation from the old set of second quantized operators $\{a_P\}$, $\{a_P^+\}$, $\{b_r\}$, $\{b_r^+\}$ to the new set of operators $\{\bar{a}_P\}$, $\{\bar{a}_P^+\}$, $\{\bar{b}_r\}$, and $\{\bar{b}_r^+\}$ has the form

$$\bar{a}_P = \sum_Q \tilde{C}_{PQ}(\tilde{B}) a_Q \tag{7.7}$$

$$\bar{a}_P^+ = \sum_Q \tilde{C}_{PQ}(\tilde{B})^+ a_Q^+ \tag{7.8}$$

$$\bar{b}_r = b_r + \sum_{PQ} \tilde{d}_{rPQ}(\tilde{B}) a_P^+ a_Q \tag{7.9}$$

$$\bar{b}_r^+ = b_r^+ + \sum_{PQ} \tilde{d}_{rPQ}(\tilde{B})^+ a_P^+ a_Q \tag{7.10}$$

The coefficients $\tilde{C}_{PQ}$ and $\tilde{d}_{rPQ}$ are given as the Taylor expansion

$$\tilde{C}_{PQ}(\tilde{B}) = \sum_{k=0}^{\infty} \frac{1}{k!} \sum_{r_1 \cdots r_k} \tilde{C}_{PQ}^{r_1 \cdots r_k} \tilde{B}_{r_1} \cdots \tilde{B}_{r_k} \tag{7.11}$$

$$\tilde{d}_{rPQ}(\tilde{B}) = \sum_{k=0} \frac{1}{k!} \sum_{s_1 \cdots s_k} \tilde{d}_{rPQ}^{s_1 \cdots s_k} \tilde{B}_{s_1} \cdots \tilde{B}_{s_k} \tag{7.12}$$

The new set of fermion operators, (7.7) and (7.8), must fulfill the fermion anticommutation relations in (3.10), and the new set of boson operators in (7.9) and (7.10) must satisfy the boson commutation rule in (3.20). This gives us some restrictive conditions on the coefficients $\tilde{C}_{PQ}$ and $\tilde{d}_{rPQ}$ (as in the case of adiabatic transformation):

$$\sum_R \tilde{C}_{PR}\tilde{C}_{QR}^+ = \delta_{PQ} \tag{7.13}$$

$$\tilde{d}_{rPQ} = \sum_R \tilde{C}_{RP}^+[b_r, \tilde{C}_{RQ}] \tag{7.14}$$

$$\tilde{d}_{rPQ} - \tilde{d}_{rPQ}^+ = 0 \tag{7.15}$$

For our purposes it is sufficient to express the Taylor expansion of equations (7.13)–(7.15) to the second order of expansion:

$$\tilde{C}_{PQ}^0 = \delta_{PQ} \tag{7.16}$$

$$\tilde{C}_{PQ}^r - \tilde{C}_{QP}^{r*} = 0 \tag{7.17}$$

$$\tilde{C}_{PQ}^{rs} + \tilde{C}_{QP}^{rs*} = \sum_R (\tilde{C}_{PR}^r \tilde{C}_{QR}^{s*} + \tilde{C}_{PR}^r \tilde{C}_{QR}^{r*}) \tag{7.18}$$

$$\tilde{d}_{rPQ}^0 = -\tilde{C}_{PQ}^r \tag{7.19}$$

$$\tilde{d}_{rPQ}^s = -\tilde{C}_{PQ}^{rs} + \sum_R \tilde{C}_{RP}^{s*}\tilde{C}_{RQ}^r \tag{7.20}$$

Similarly, as in the case of adiabatic transformation, we have two invariants of the new nonadiabatic transformation. The number of fermion operators is the invariant of the transformation

$$\bar{N} = \sum_P \bar{a}_P^+ \bar{a}_P = \sum_P a_P^+ a_P = N \tag{7.21}$$

and the momentum operator $\tilde{B}_r$ is also the invariant of the transformation

$$\tilde{\tilde{B}} = \tilde{B}_r \tag{7.22}$$

In analogy with the adiabatic transformation it is easy to show the group property of the new nonadiabatic transformations. Therefore, there exists the inverse transformation from the new set of second quantized operators to the old set of second quantized operators

$$a_P = \sum_Q \tilde{\tilde{C}}_{PQ}(\bar{\tilde{B}})\bar{a}_Q \tag{7.23}$$

$$b_r = \bar{b}_r + \sum_{PQ} \tilde{\tilde{d}}_{rPQ}(\bar{\tilde{B}})\bar{a}_P^+ \bar{a}_Q \tag{7.24}$$

and, similarly, for the creation operators a_P^+ and b_r^+. The coefficients $\tilde{\tilde{C}}_{PQ}$ and $\tilde{\tilde{d}}_{rPQ}$ are given through the coefficients C_{PQ} and d_{rPQ} as

$$\tilde{\tilde{C}}_{PQ}(\bar{\tilde{B}}) = \tilde{\tilde{C}}_{PQ}(\tilde{B}) = \bar{C}_{QP}(\tilde{B})^+ \tag{7.25}$$

$$\tilde{\tilde{d}}_{rPQ}(\bar{\tilde{B}}) = \tilde{\tilde{d}}_{rPQ}(\tilde{B}) = -\sum_{RS} \tilde{C}_{PR}(\tilde{B})\tilde{d}_{rRS}(\tilde{B})\tilde{C}_{QS}(\tilde{B})^+ \tag{7.26}$$

Because we work with three representations of the V–E Hamiltonian, namely, with crude, adiabatic, and nonadiabatic representations, and with two transformations, that is, adiabatic and nonadiabatic, we introduce the new notation. We denote the second quantized operators in crude representation for electrons as $\{\bar{a}_P\}$, $\{\bar{a}_P^+\}$; for phonons as $\{\bar{b}_r\}$, $\{\bar{b}_r^+\}$; for coordinate operators as $\{\bar{B}_r\}$; and for momentum operators as $\{\bar{\tilde{B}}_r\}$.

The total V–E Hamiltonian has in the crude representation the form (2.1)

$$H = T_N(\bar{\tilde{B}}) + E_{NN}(\bar{B}) + \sum_{PQ} h_{PQ}(\bar{B})\bar{a}_P^+ \bar{a}_Q + \tfrac{1}{2} \sum_{PQRS} v_{PQRS}^0 \bar{a}_P^+ \bar{a}_Q^+ \bar{a}_S \bar{a}_R \tag{7.27}$$

The quantities $E_{NN}(\bar{B})$ and $h_{PQ}(\bar{B})$ are given through the Taylor expansion of (2.12) and (2.13) as

$$E_{NN}(\bar{B}) = \sum_{i=0}^{\infty} E_{NN}^{(i)}(\bar{B}) \tag{7.28}$$

and

$$h_{PQ}(\bar{B}) = h_{PQ}^{0} + \sum_{i=1}^{\infty} u_{PQ}^{(i)}(\bar{B}) \tag{7.29}$$

where

$$u_{PQ}(R) = \left\langle P \left| \sum_j \frac{-Z_j e^2}{|\mathbf{r} - \mathbf{R}_j|} \right| Q \right\rangle \tag{7.30}$$

After introducing the normal coordinates we are able to define the vibrational Hamiltonian as

$$H_B = \sum_r \hbar\omega_r(\bar{b}_r^+ \bar{b}_r + 1/2) \tag{7.31}$$

The crucial step in the breakdown of the Born–Oppenheimer approximation is the quantization of the vibrational Hamiltonian H_B, which contains the kinetic energy and also the potential energy

$$H_B = E_{\text{kin}}(\bar{B}) + E_{\text{pot}}(\bar{B}) \tag{7.32}$$

The potential energy is determined by the quadratic part of the nuclear energy $E_{NN}^{(2)}(\bar{B})$ as well as by some potential energy $V_N^{(2)}(\bar{B})$ which is also the quadratic function of coordinate operators $\bar{B}_r$ and has its origin in the interaction of the electrons with the vibrating nuclei. Therefore, we have

$$E_{\text{pot}}(\bar{B}) = E_{NN}^{(2)}(\bar{B}) + V_N^{(2)}(\bar{B}) \tag{7.33}$$

In the case of the kinetic energy term we think this is identical with the kinetic energy of the nuclei. This was true for the case of adiabatic approximation. In the case of the breakdown of adiabatic approximation we have to remember the finite mass of electrons and therefore to introduce the more general kinetic energy term. Therefore, we add to the kinetic energy of the nuclei $T_N(\tilde{\bar{B}})$ some other yet unknown term which we denote $W_N^{(2)}(\tilde{\bar{B}})$ and which will be the quadratic function of momentum operator

$$E_{\text{kin}}(\tilde{\bar{B}}) = T_N(\tilde{\bar{B}}) + W_N^{(2)}(\tilde{\bar{B}}) \tag{7.34}$$

The total V–E Hamiltonian (7.27) can be divided into two parts,

$$H = H_A + H_B \tag{7.35}$$

where H_B is the vibrational Hamiltonian and from equations (7.33) and (7.34) has the form

$$H_B = T_N(\tilde{\bar{B}}) + W_N^{(2)}(\tilde{\bar{B}}) + E_{NN}^{(2)}(\bar{B}) + V_N^{(2)}(\bar{B}) \tag{7.36}$$

The Hamiltonian H_A therefore is

$$\begin{aligned} H_A = {} & E_{NN}(\bar{B}) - E_{NN}^{(2)}(\bar{B}) - V_N^{(2)}(\bar{B}) - W_N^{(2)}(\bar{B}) \\ & + \sum_{PQ} h_{PQ}(\bar{B})\bar{a}_P^+\bar{a}_Q + \tfrac{1}{2} \sum_{PQRS} v_{PQRS}^0 \bar{a}_P^+\bar{a}_Q^+\bar{a}_S\bar{a}_R \end{aligned} \tag{7.37}$$

Further, we limit ourselves to such an approximation that the Hamiltonian in (7.35) in nonadiabatic representation will be given up to the second order in boson indices. In individual orders of the Taylor expression we use the notation $H_{A\bar{i}(j,k)}$, where $\bar{i}$ means the order of the Taylor expansion in the crude representation, j means the power of the coordinate operator $\bar{B}$, and k means the power of the momentum operator $\tilde{\bar{B}}$. Therefore, we have

$$H_{A\bar{0}} = E_{NN}^0 + \sum_{PQ} h_{PQ}^0 \bar{a}_P^+ \bar{a}_Q + \tfrac{1}{2} \sum_{PQRS} v_{PQRS}^0 \bar{a}_P^+\bar{a}_Q^+\bar{a}_S\bar{a}_R \tag{7.38}$$

$$H_{A\bar{1}(1,0)} = \sum_r E_{NN}^r \bar{B}_r + \sum_{PQr} u_{PQ}^r \bar{B}_r \bar{a}_P^+\bar{a}_Q \tag{7.39}$$

$$H_{A\bar{2}(2,0)} = -\tfrac{1}{2} \sum_{rs} V_N^{rs} \bar{B}_r\bar{B}_s + \tfrac{1}{2} \sum_{PQrs} u_{PQ}^{rs} \bar{B}_r\bar{B}_s\bar{a}_P^+\bar{a}_Q \tag{7.40}$$

$$H_{A\bar{2}(2,0)} = -\tfrac{1}{2} \sum_{rs} W_N^{rs} \tilde{\bar{B}}_r\tilde{\bar{B}}_s \tag{7.41}$$

We also try to determine the quantities V_N^{rs} and W_N^{rs} in order to get the vibrational frequencies ω_r in the case of the breakdown of the Born–Oppenheimer approximation.

Until now, the whole problem has been formulated only in the crude representation. The final result we get after passing to the nonadiabatic representation. First, we have to pass from the crude representation to the adiabatic representation. Let us denote the fermion operators in adiabatic representation as $\{\bar{\bar{a}}_P\}$, $\{\bar{\bar{a}}_P^+\}$; the boson operators as $\{\bar{\bar{b}}_r\}$, $\{\bar{\bar{b}}_r^+\}$; the coordinate operators as $\{\bar{\bar{B}}_r\}$; and momentum operators as $\{\bar{\bar{\bar{B}}}\}$. The adiabatic transformation in our new notation is then

$$\bar{a}_P = \sum_Q C_{PQ}(\bar{\bar{B}})\bar{\bar{a}}_Q \tag{7.42}$$

$$\bar{a}_P^+ = \sum_Q C_{PQ}(\bar{\bar{B}})^+\bar{\bar{a}}_Q^+ \tag{7.43}$$

$$\bar{b}_r = \bar{\bar{b}}_r + \sum_{PQ} d_{rPQ}(\bar{\bar{B}})\bar{\bar{a}}_P^+\bar{\bar{a}}_Q \tag{7.44}$$

$$\bar{b}_r^+ = \bar{\bar{b}}_r^+ - \sum_{PQ} d_{rPQ}(\bar{\bar{B}})\bar{\bar{a}}_P^+\bar{\bar{a}}_Q \tag{7.45}$$

If we apply this transformation to the V–E Hamiltonian H in (7.35) we get for the individual orders of the Taylor expansion terms $H_{A\bar{\bar{i}}(j,k)}$ and $H_{B\bar{\bar{i}}(j,k)}$ where $\bar{\bar{i}}$ is the order of the Taylor expansion in the adiabatic representation, j means the power of the coordinate operator $\bar{\bar{B}}$, and k means the power of the momentum operator $\bar{\bar{\bar{B}}}$. Expanding equations (7.38)–(7.41) and (7.31) in adiabatic representation, we get

$$H_{A\bar{0}} \rightarrow H_{A\bar{\bar{0}}} + H_{A\bar{\bar{1}}(1,0)} + H_{A\bar{\bar{2}}(2,0)} \tag{7.46}$$

$$H_{A\bar{1}(1,0)} \rightarrow H_{A\bar{\bar{1}}(1,0)} + H_{A\bar{\bar{2}}(2,0)} \tag{7.47}$$

$$H_{A\bar{2}(2,2)} \rightarrow H_{A\bar{\bar{2}}(2,0)} \tag{7.48}$$

$$H_{A\bar{2}(0,2)} \rightarrow H_{A\bar{\bar{2}}(0,2)} \tag{7.49}$$

$$H_B = H_{B\bar{0}} \rightarrow H_{B\bar{\bar{1}}} + H_{B\bar{\bar{1}}(0,1)} + H_{B\bar{\bar{2}}(1,1)} + H_{B\bar{\bar{2}}(0,0)} \tag{7.50}$$

where

$$H_{A\bar{\bar{0}}} = E_{NN}^0 + \sum_{PQ} h_{PQ}^0\bar{\bar{a}}_P^+\bar{\bar{a}}_Q + \tfrac{1}{2}\sum_{PQRS} v_{PQRS}^0\bar{\bar{a}}_P^+\bar{\bar{a}}_Q^+\bar{\bar{a}}_S\bar{\bar{a}}_R \tag{7.51}$$

$$\begin{aligned} H_{A\bar{\bar{1}}(1,0)} = {} & \sum_r E_{NN}^r\bar{\bar{B}}_r + \sum_{PQr}\left[u_{PQ}^r + \sum_R (h_{PR}^0 C_{RQ}^r + h_{RQ}^0 C_{RP}^{r*})\right]\bar{\bar{B}}_r\bar{\bar{a}}_P^+\bar{\bar{a}}_Q \\ & + \sum_{PQRSTr} (v_{PQTS}^0 C_{TR}^r + v_{TQRS}^0 C_{PT}^{r*})\bar{\bar{B}}_r\bar{\bar{a}}_P^+\bar{\bar{a}}_Q^+\bar{\bar{a}}_S\bar{\bar{a}}_R \end{aligned} \tag{7.52}$$

$$H_{A\bar{\bar{2}}(2,0)} = -\tfrac{1}{2}\sum_{rs} V_N^{rs}\bar{B}_r\bar{B}_s + \sum_{PQrs}\Bigg[\tfrac{1}{2}u_{PQ}^{rs} + \sum_R(\tfrac{1}{2}h_{PR}^0 C_{RQ}^{rs} + \tfrac{1}{2}h_{RQ}^0 C_{RP}^{rs*}$$

$$+ u_{PR}^r C_{RQ}^s + u_{RQ}^r C_{RP}^{s*}) + \sum_{RS} h_{RS}^0 C_{RP}^{r*} C_{SQ}^s\Bigg]\bar{B}_r\bar{B}_s\bar{a}_P^+\bar{a}_Q$$

$$+ \tfrac{1}{2}\sum_{PQRSTrs}\Bigg\{v_{PQTS}^0 C_{TR}^{rs} + v_{TQRS}^0 C_{TP}^{rs*} + \sum_U [v_{PQTU}^0 C_{TR}^r C_{US}^s$$

$$+ v_{TURS}^0 C_{TP}^{r*} C_{UQ}^{s*} + 2(v_{TQUS}^0 - v_{TQSU}^0)C_{TP}^{r*}C_{UR}^s]\Bigg\}\bar{B}_r\bar{B}_s\bar{a}_P^+\bar{a}_Q^+\bar{a}_S\bar{a}_R$$

(7.53)

$$H_{A\bar{\bar{2}}(0,2)} = -\tfrac{1}{2}\sum_{rs} W_N^{rs}\bar{\tilde{B}}_r\bar{\tilde{B}}_s \tag{7.54}$$

$$H_{B\bar{\bar{0}}} = \sum_r \hbar\omega_r(\bar{b}_r^+\bar{b}_r + 1/2) \tag{7.55}$$

$$H_{B\bar{\bar{1}}(0,1)} = -\sum_{PQr}\hbar\omega_r d_{rPQ}^0\bar{\tilde{B}}_r\bar{a}_P^+\bar{a}_Q \tag{7.56}$$

$$H_{B\bar{\bar{2}}(1,1)} = -\tfrac{1}{2}\sum_{PQrs}\hbar\omega_r d_{rPQ}^s(\bar{\tilde{B}}_r\bar{\tilde{B}}_s + \bar{\tilde{B}}_s\bar{\tilde{B}}_r)\bar{a}_P^+\bar{a}_Q \tag{7.57}$$

$$H_{B\bar{\bar{2}}(0,0)} = -\sum_{PQRr}\hbar\omega_r d_{rPR}^0 d_{rRQ}^0\bar{a}_P^+\bar{a}_Q - \sum_{PQRSr}\hbar\omega_r d_{rPR}^0 d_{rQS}^0\bar{a}_P^+\bar{a}_Q^+\bar{a}_S\bar{a}_R \tag{7.58}$$

The final step (the breakdown of the Born–Oppenheimer approximation) will be in passing from the adiabatic representation to the nonadiabatic representation using the nonadiabatic transformation. Let us denote the fermion operators in nonadiabatic representation as $\{a_P\}$, $\{a_P^+\}$; the boson operators as $\{b_r\}$, $\{b_r^+\}$; the coordinate operator as $\{B_r\}$; and the momentum operators as $\{\tilde{B}_r\}$. The nonadiabatic transformation in the new notation then reads

$$\bar{a}_P = \sum_Q \tilde{C}_{PQ}(\tilde{B})a_Q \tag{7.59}$$

$$\bar{a}_P^+ = \sum_Q \tilde{C}_{PQ}(\tilde{B})^+ a_Q^+ \tag{7.60}$$

$$\bar{b}_r = b_r + \sum_{PQ}\tilde{d}_{rPQ}(\tilde{B})d_P^+ a_Q \tag{7.61}$$

$$\bar{b}_r^+ = b_r^+ + \sum_{PQ}\tilde{d}_{rPQ}(\tilde{B})a_P^+ a_Q \tag{7.62}$$

Up to the second order of the Taylor expansion the terms $H_{A\bar{\bar{i}}(j,k)}$ and $H_{B\bar{\bar{i}}(j,k)}$ will change to the terms $H_{A_i(j,k)}$ and $H_{B_i(j,k)}$, respectively, according to

$$H_{A\bar{\bar{0}}} \to H_{A_0} + H_{A_{1(0,1)}} + H_{A_{2(0,2)}} \tag{7.63}$$

$$H_{A\bar{\bar{1}}(1,0)} \to H_{A_{1(1,0)}} + H_{A_{2(1,1)}} + H_{A_{2(0,0)}} \tag{7.64}$$

$$H_{A\bar{\bar{2}}(2,0)} \to H_{A_{2(2,0)}} \tag{7.65}$$

$$H_{A\bar{\bar{2}}(0,2)} \to H_{A_{2(2,0)}} \tag{7.66}$$

$$H_{B\bar{\bar{0}}} \to H_{B_0} + H_{B_1(1,0)} + H_{B_{2(1,1)}} + H_{B_{2(0,0)}} \tag{7.67}$$

$$H_{B\bar{\bar{1}}(0,1)} \to H_{B_{1(0,1)}} + H_{B_{2(0,2)}} \tag{7.68}$$

$$H_{B\bar{\bar{2}}(1,1)} \to H_{B_{2(1,1)}} \tag{7.69}$$

$$H_{B\bar{\bar{2}}(0,0)} \to H_{B_{2(0,0)}} \tag{7.70}$$

The terms $A_{A_i(j,k)}$ and $H_{B_i(j,k)}$ will have the form

$$H_{A_0} = E^0_{NN} + \sum_{PQ} h^0_{PQ} a^+_P a_Q + \tfrac{1}{2} \sum_{PQRS} v^0_{PQRS} a^+_P a^+_Q a_S a_R \tag{7.71}$$

$$\begin{aligned} H_{A_{1(1,0)}} = {} & \sum_r E^r_{NN} B_r + \sum_{PQr} \left[u^r_{PQ} + \sum_R (h^0_{PR} C^r_{RQ} + h^0_{RQ} C^{r*}_{RP}) \right] B_r a^+_P a_Q \\ & + \sum_{PQRSTr} (v^0_{PQTS} C^r_{TR} + v^0_{TQRS} C^{r*}_{TP}) B_r a^+_P a^+_Q a_S a_R \end{aligned} \tag{7.72}$$

$$\begin{aligned} H_{A_{1(0,1)}} = {} & \sum_{PQRr} (h^0_{PR} \tilde{C}^r_{RQ} - h^0_{RQ} \tilde{C}^{r*}_{RP}) \tilde{B}_r a^+_P a_Q \\ & + \sum_{PQRSTr} (v^0_{PQTS} \tilde{C}^r_{TR} - v^0_{TQRS} \tilde{C}^{r*}_{TP}) \tilde{B}_r a^+_P a^+_Q a_S a_R \end{aligned} \tag{7.73}$$

$$\begin{aligned} H_{A_{2(1,1)}} = {} & \tfrac{1}{2} \sum_{PQRrs} \Big\{ u^r_{PR} \tilde{C}^s_{RQ} - u^r_{RQ} \tilde{C}^{s*}_{RP} + \sum_S [(h^0_{PR} C^r_{RS} + h^0_{RS} C^{r*}_{RP}) \tilde{C}^s_{SQ} \\ & - (h^0_{SR} C^r_{RQ} + h^0_{RQ} C^{r*}_{RS}) \tilde{C}^{s*}_{SP}] \Big\} (B_r \tilde{B}_s + \tilde{B}_s B_r) a^+_P a_Q \\ & + \tfrac{1}{2} \sum_{PQRSTU} [(v^0_{PQTS} C^r_{TU} - v^0_{PQTU} C^r_{TS}) \tilde{C}^s_{UR} \\ & + (v^0_{TURS} C^{r*}_{TQ} - v^0_{TQRS} C^{r*}_{TU}) \\ & \cdot \tilde{C}^{s*}_{UP} + (v^0_{TQUS} - v^0_{TQSU})(C^{r*}_{TP} \tilde{C}^s_{UR} - C^r_{UR} \tilde{C}^{s*}_{TP})] \\ & \cdot (B_r \tilde{B}_s + \tilde{B}_s B_r) a^+_P a^+_Q a_S a_R \end{aligned} \tag{7.74}$$

$$
\begin{aligned}
H_{A_{2(2,0)}} = & -\tfrac{1}{2}\sum_{rs} V_N^{rs} B_r B_s + \sum_{PQrs}\Bigg[\tfrac{1}{2}u_{PQ}^{rs} + \sum_{R}\left(\tfrac{1}{2}h_{PR}^{0} C_{RQ}^{rs} + \tfrac{1}{2}h_{RQ}^{0} C_{RP}^{rs*}\right.\\
& \left. + u_{PR}^{r} C_{RQ}^{s} + u_{RQ}^{r} C_{RP}^{s*}\right) + \sum_{RS} h_{RS}^{0} C_{RP}^{r*} C_{SQ}^{s}\Bigg] B_r B_s a_P^{+} a_Q \\
& + \tfrac{1}{2}\sum_{PQRSTrs}\Bigg\{ v_{PQTS}^{0} C_{TR}^{rs} + v_{TQRS}^{0} C_{TP}^{rs*} \\
& + \sum_{U}\left[v_{PQTU}^{0} C_{TR}^{r} C_{US}^{s} + v_{TURS}^{0} C_{TP}^{r*} C_{UQ}^{s*} \right.\\
& \left. + 2\left(v_{TQUS}^{0} - v_{TQSU}^{0}\right) C_{TP}^{r*} C_{UR}^{s}\right]\Bigg\} B_r B_s a_P^{+} a_Q^{+} a_S a_R
\end{aligned}
\tag{7.75}
$$

$$
\begin{aligned}
H_{A_{2(0,2)}} = & -\tfrac{1}{2}\sum_{rs} W_N^{rs} \tilde{B}_r \tilde{B}_s + \sum_{PQRrs}\Bigg(\tfrac{1}{2}h_{PR}^{0} \tilde{C}_{RQ}^{rs} \\
& + \tfrac{1}{2}h_{RQ}^{0} \tilde{C}_{RP}^{rs*} - \sum_{S} h_{RS}^{0} \tilde{C}_{RP}^{r*} \tilde{C}_{SQ}^{s}\Bigg) \tilde{B}_r \tilde{B}_s a_P^{+} a_Q \\
& + \tfrac{1}{2}\sum_{PQRSTrs}\Bigg\{ v_{PQTS}^{0} \tilde{C}_{TR}^{rs} + v_{TQRS}^{0} \tilde{C}_{TP}^{rs*} \\
& + \sum_{U}\left[v_{PQTU}^{0} \tilde{C}_{TR}^{r} \tilde{C}_{US}^{s} + v_{TURS}^{0} \tilde{C}_{TP}^{r*} \tilde{C}_{UQ}^{s*} \right.\\
& \left. + 2\left(v_{TQSU}^{0} - v_{TQUS}^{0}\right) \tilde{C}_{TP}^{r*} \tilde{C}_{UR}^{s}\right]\Bigg\} \tilde{B}_r \tilde{B}_s a_P^{+} a_Q^{+} a_S a_R
\end{aligned}
\tag{7.76}
$$

$$
\begin{aligned}
H_{A_{2(0,0)}} = & \; 2\sum_{PQr} E_{NN}^{r} \tilde{d}_{rPQ}^{0} a_P^{+} a_Q \\
& + \sum_{PQRr}\Bigg\{\Bigg[u_{PR}^{r} + \sum_{S}\left(h_{PS}^{0} C_{SR}^{r} + h_{SR}^{0} C_{SP}^{r*}\right)\Bigg] \tilde{d}_{rRQ}^{0} \\
& + \Bigg[u_{RQ}^{r} + \sum_{S}\left(h_{RS}^{0} C_{SQ}^{r} + h_{SQ}^{0} C_{SR}^{r*}\right)\Bigg] \tilde{d}_{rPR}^{0}\Bigg\} a_P^{+} a_Q \\
& + 2\sum_{PQRSr}\Bigg[u_{PR}^{r} + \sum_{T}\left(h_{PT}^{0} C_{TR}^{r} + h_{TR}^{0} C_{TP}^{r*}\right)\Bigg] \tilde{d}_{rQS}^{0} a_P^{+} a_Q^{+} a_S a_R \\
& + \sum_{PQRSTUr}\left[\left(v_{PQTS}^{0} C_{TU}^{r} - v_{PQTU}^{0} C_{TS}^{r}\right)\tilde{d}_{rUR}^{0}\right. \\
& + \left(v_{TQRS}^{0} C_{TU}^{r*} - v_{TURS}^{0} C_{TQ}^{r*}\right)\tilde{d}_{rPU}^{0}
\end{aligned}
$$

$$+ (v^0_{TQUS} - v^0_{TQSU})(C^{r*}_{TP}\bar{d}^0_{rUR} + C^r_{UR}\bar{d}^0_{rPT})]a^+_P a^+_Q a_S a_R$$

$$+ 2 \sum_{PQRSTUVr} (v^0_{PQVT}C^r_{VS} + v^0_{VQST}C^{r*}_{VP})\bar{d}^0_{rRU}a^+_P a^+_Q a^+_R a_U a_T a_S$$

$$(7.77)$$

$$H_{B_0} = \sum_r \hbar\omega_r(b^+_r b_r + 1/2) \tag{7.78}$$

$$H_{B_{1(1,0)}} = \sum_{PQr} \hbar\omega_r \bar{d}^0_{rPQ} B_r a^+_P a_Q \tag{7.79}$$

$$H_{B_{1(0,1)}} = -\sum_{PQr} \hbar\omega_r d^0_{rPQ} \tilde{B}_r a^+_P a_Q \tag{7.80}$$

$$H_{B_{2(1,1)}} = \tfrac{1}{2} \sum_{PQrs} (\hbar\omega_r \bar{d}^s_{rPQ} - \hbar\omega_s d^r_{sPQ})(B_r\tilde{B}_s + \tilde{B}_s B_r)a^+_P a_Q \tag{7.81}$$

$$H_{B_{2(2,0)}} = 0 \tag{7.82}$$

$$H_{B_{2(0,2)}} = -\sum_{PQRrs} \hbar\omega_r(d^0_{rPR}\tilde{C}^s_{RQ} - d^0_{rRQ}\tilde{C}^{s*}_{RP})\tilde{B}_r\tilde{B}_s a^+_P a_Q \tag{7.83}$$

$$H_{B_{2(0,0)}} = \sum_{PQRr} \hbar\omega_r(\bar{d}^0_{rPR}\bar{d}^0_{rRQ} - d^0_{rPR}d^0_{rRQ})a^+_P a_Q$$

$$+ \sum_{PQRSr} \hbar\omega_r(\bar{d}^0_{rPR}\bar{d}^0_{rQS} - d^0_{rPR}d^0_{rQS})a^+_P a^+_Q a_S a_R \tag{7.84}$$

As we have shown, both the adiabatic and the nonadiabatic transformations preserve the number of fermions. Therefore, we can apply the Wick theorem to equations (7.71)–(7.84). For the fermions we use the indices P, Q, R, S, T, U, V. Further, for occupied states we use the indices A, B, C, D and for the unoccupied states the indices I, J, K, L. The one-fermion terms we write in normal product form according to

$$\sum_{PQ} \lambda_{PQ} a^+_P a_Q = \sum_{PQ} \lambda_{PQ} N[a^+_P a_Q] + \lambda \tag{7.85}$$

where

$$\lambda = \sum_I \lambda_{II} \tag{7.86}$$

Analogously, for two-fermion terms we have

$$\sum_{PQRS} \mu_{PQRS} a^+_P a^+_Q a_S a_R$$

$$= \sum_{PQRS} \mu_{PQRS} N[a^+_P a^+_Q a_S a_R] + \sum_{PQ} \mu_{PQ} N[a^+_P a_Q] + \mu \tag{7.87}$$

where

$$\mu_{PQ} = \sum_{I} (\mu_{PIQI} + \mu_{IPIQ} - \mu_{PIIQ} - \mu_{IPQI}) \tag{7.88}$$

and

$$\mu = \sum_{IJ} (\mu_{IJIJ} - \mu_{IJJI}) \tag{7.89}$$

Finally, for the three-fermion terms

$$\begin{aligned} &\sum_{PQRSTU} v_{PQRSTU} a_P^+ a_Q^+ a_R^+ a_U a_T a_S \\ &\quad = \sum_{PQRSTU} v_{PQRSTU} N[a_P^+ a_Q^+ a_R^+ a_U a_T a_S] + \sum_{PQRS} v_{PQRS} N[a_P^+ a_Q^+ a_S a_R] \\ &\quad + \sum_{PQ} v_{PQ} N[a_P^+ a_Q] + v \end{aligned} \tag{7.90}$$

where

$$\begin{aligned} v_{PQRS} = \sum_{I} (&v_{PQIRSI} - v_{PQIRIS} + v_{PQIIRS} - v_{PIQRSI} \\ &+ v_{PIQRIS} - v_{PIQIRS} + v_{IPQRSI} - v_{IPQRIS} + v_{IPQIRS}) \end{aligned} \tag{7.91}$$

$$\begin{aligned} v_{PQ} = \sum_{IJ} (&v_{PIJQIJ} - v_{PIJQJI} - v_{PIJIQJ} + v_{PIJJQI} + v_{PIJIJQ} - v_{PIJJIQ} \\ &- v_{IPJQIJ} + v_{IPJQJI} + v_{IPJIQJ} - v_{IPJJQI} - v_{IPJIJQ} \\ &+ v_{IPJJIQ} + v_{IJPQIJ} - v_{IJPQJI} \\ &- v_{IJPIQJ} + v_{IJPJQI} + v_{IJPIJQ} - v_{IJPJIQ}) \end{aligned} \tag{7.92}$$

and

$$v = \sum_{IJK} (v_{IJKIJK} - v_{IJKIKJ} - v_{IJKJIK} + v_{IJKJKI} + v_{IJKKIJ} - v_{IJKKJI}) \tag{7.93}$$

We denote the individual terms of the nonadiabatic Hamiltonian that do not contain the fermion operators as H^0; the terms that contain the normal product $N[a_P^+ a_Q]$ as H'; the two-fermion terms that contain the normal product $N[a_P^+ a_Q^+ a_S a_R]$ as H''; and, finally, the three-fermion terms that contain the normal product $N[a_P^+ a_Q^+ a_R^+ a_U a_T a_S]$ as H'''.

We also try to obtain the equations for the new unperturbed fermion energies and for the coefficients of the nonadiabatic transformation. We proceed in a similar manner to the case of adiabatic representation where the new fermion energies were obtained by solving the *CPHF* equations. Therefore, analogously, we want to diagonalize the one-fermion terms H' in the nonadiabatic representation; that is, we want to have

$$\mathbf{H}' = \sum_P X_P N[a_P^+ a_P] \tag{7.94}$$

We try to diagonalize equation (7.94) to the first order of the Taylor expansion. In the zeroeth order according to (7.71) we have

$$H_0' = H_{A_0'} = \sum_{PQ} \left[h_{PQ}^0 + \sum_I (v_{PIQI}^0 - v_{PIIQ}^0) \right] N[a_P^+ a_Q] = \sum_P \epsilon_P^0 N[a_P^+ a_P] \tag{7.95}$$

which is the diagonalization of the Hartree–Fock operator

$$f_{PQ}^0 = h_{PQ}^0 + \sum_I (v_{PIQI}^0 - v_{PIIQ}^0) = \epsilon_P^0 \, \delta_{PQ} \tag{7.96}$$

The zeroeth order is therefore the same as in the crude as well as in adiabatic representation. In the first order of the Taylor expansion we get, according to (7.72), (7.73), (7.79), and (7.80),

$$\begin{aligned} H_{A_1'(1,0)} = \sum_{PQr} \Big\{ & u_{PQ}^r + (\epsilon_P^0 - \epsilon_Q^0) C_{PQ}^r \\ & + \sum_{AI} [(v_{PIQA}^0 - v_{PIAQ}^0) C_{AI}^r - (v_{PAQI}^0 - v_{PAIQ}^0) C_{IA}^r] \Big\} B_r N[a_P^+ a_Q] \end{aligned} \tag{7.97}$$

$$\begin{aligned} H_{A_1'(1,0)} = \sum_{PQr} \Big\{ & (\epsilon_P^0 - \epsilon_Q^0) \tilde{C}_{PQ}^r \\ & + \sum_{AI} [(v_{PIQA}^0 - v_{PIAQ}^0) \tilde{C}_{AI}^r - (v_{PAQI}^0 - v_{PAIQ}^0) \tilde{C}_{IA}^r] \Big\} \tilde{B}_r N[a_P^+ a_Q] \end{aligned} \tag{7.98}$$

$$H_{B_1'(1,0)} = - \sum_{PQr} \hbar\omega_r \tilde{C}_{PQ}^r B_r N[a_P^+ a_Q] \tag{7.99}$$

$$H_{B_1'(0,1)} = - \sum_{PQr} \hbar\omega_r C_{PQ}^r \tilde{B}_r N[a_P^+ a_Q] \tag{7.100}$$

If we diagonalize the terms $H'_{1(1,0)}$ and $H'_{1(0,1)}$ we get

$$H'_{1(1,0)} = \sum_{Pr} \epsilon^r_P B_r N[a^+_P a_P] \tag{7.101}$$

and

$$H'_{1(0,1)} = \sum_{Pr} \tilde{\epsilon}^r_P \tilde{B}_r N[a^+_P a_P] \tag{7.102}$$

which leads to the coupled equations for C^r_{PQ} and $\tilde{C}^r_{PQ}$ coefficients

$$\begin{aligned} u^r_{PQ} + (\epsilon^0_P - \epsilon^0_Q)C^r_{PQ} + \sum_{AI} [(v^0_{PIQA} - v^0_{PIAQ})C^r_{AI} \\ - (v^0_{PAQI} - v^0_{PAIQ})C^r_{IA}] - \hbar\omega_r \tilde{C}^r_{PQ} = \epsilon^r_P \, \delta_{PQ} \end{aligned} \tag{7.103}$$

and

$$\begin{aligned} (\epsilon^0_P - \epsilon^0_Q)\tilde{C}^r_{PQ} + \sum_{AI} [(v^0_{PIQA} - v^0_{PIAQ})\tilde{C}^r_{AI} \\ - (v^0_{PAQI} - v^0_{PAIQ})\tilde{C}^r_{IA}] - \hbar\omega_r C^r_{PQ} = \tilde{\epsilon}^r_P \, \delta_{PQ} \end{aligned} \tag{7.104}$$

Equations (7.103) and (7.104) differ from the analogous equations in the crude and in the adiabatic representation. For example, in the crude representation for which $C = 0$ and $\tilde{C} = 0$, we get the nonphysical equation

$$u^r_{PQ} = \epsilon^r_P \, \delta_{PQ} \tag{7.105}$$

which of course cannot be diagonalized. In adiabatic representation for which $\tilde{C} = 0$, we get

$$\begin{aligned} u^0_{PQ} + (\epsilon^0_P - \epsilon^0_Q)C^r_{PQ} + \sum_{AI} [(v^0_{PIQA} - v^0_{PIAQ})C^r_{AI} \\ - (v^0_{PAQI} - v^0_{PAIQ})C^r_{IA}] = \epsilon^r_P \, \delta_{PQ} \end{aligned} \tag{7.106}$$

and

$$-\hbar\omega_r C^r_{PQ} = \tilde{\epsilon}^r_P \, \delta_{PQ} \tag{7.107}$$

Here we cannot diagonalize equation (7.107). We can omit equation (7.107) under the condition that

$$\hbar\omega_r|C^r_{PQ}| \ll u^r_{PQ} \sim |\epsilon^0_P - \epsilon^0_Q|\,|C^r_{PQ}| \tag{7.108}$$

which is equivalent to the condition

$$\hbar\omega_r \ll |\epsilon^0_P - \epsilon^0_Q| \tag{7.109}$$

Condition (7.109) is nothing else than the condition for the validity of the adiabatic approximation. Therefore, in the adiabatic representation we perform the diagonalization according to (7.106). When condition (7.108) is not satisfied we have to perform the diagonalization according to equations (7.103) and (7.104). Solving equations (7.103) and (7.104) we obtain the nondiagonal adiabatic coefficients C^r_{PQ} and the nondiagonal coefficient $\tilde{C}^r_{PQ}$, $(P \neq Q)$. In the case $P = Q$ we can simply put

$$C^r_{PP} = 0 \tag{7.110}$$

and

$$\tilde{C}^r_{PP} = 0 \tag{7.111}$$

To continue in putting the V–E nonadiabatic Hamiltonian into the diagonal form we can try to make zero those nonfermion terms that are linear or quadratic functions of the boson operators, that is,

$$\sum_i H^0_{i(j,k)} = 0, \qquad j + k \in \{1, 2\} \tag{7.112}$$

We limit ourselves to the second order of the Taylor expansion. In the first order, according to (7.72), (7.73), (7.79), (7.80), (7.110), and (7.111), we have

$$H^0_{A1(1,0)} = \sum_r \left(E^r_{NN} + \sum_I u^r_I \right) B_R \tag{7.113}$$

$$H^0_{A1(0,1)} = 0 \tag{7.114}$$

$$H^0_{B1(1,0)} = 0 \tag{7.115}$$

$$H^0_{B1(0,1)} = 0 \tag{7.116}$$

This leads to the equation

$$H^0_{A_{1(1,0)}} = \sum_r \left(E^r_{NN} + \sum_I u^r_{II} \right) B_r = 0 \tag{7.117}$$

If we denote the first order force constants as

$$E^r = E^r_{NN} + \sum_I u^r_{II} \tag{7.118}$$

we get

$$E^r = 0 \tag{7.19}$$

This is the condition for the equilibrium position of the nuclei. We see that this condition is the same in all three representations. Note that in nonadiabatic representation we would get the more exact result if we also added to the term $H^0_{1(0,1)}$ the terms originating from the third order of the Taylor expansion $H^0_{3(1,0)}$ and $H^0_{3(0,1)}$. Their derivation is rather complex, and we do not derive them here. Taking the second order of the Taylor expansion $H^0_{2(1,1)}$ and asking

$$H^0_{2(1,1)} = 0 \tag{7.120}$$

we have according to (7.74) and (7.81)

$$H^0_{2(1,1)} = \tfrac{1}{2} \sum_{AIrs} \hbar\omega_r (\tilde{C}^r_{IA}\tilde{C}^s_{AI} - \tilde{C}^r_{AI}\tilde{C}^s_{IA})(B_r\tilde{B}_s + \tilde{B}_sB_r) \tag{7.121}$$

and

$$H^0_{B_{2(1,1)}} = \tfrac{1}{4} \sum_{Irs} \left\{ \hbar\omega_r \left[\tilde{C}^{rs*}_{II} - \tilde{C}^{rs}_{II} - \sum_A (\tilde{C}^r_{IA}\tilde{C}^s_{AI} - \tilde{C}^r_{AI}\tilde{C}^s_{IA}) \right] \right.$$
$$\left. + \hbar\omega_s \left[C^{rs*}_{II} - C^{rs}_{II} + \sum_A (C^r_{IA}C^s_{AI} - C^r_{AI}C^s_{IA}) \right] \right\}(B_r\tilde{B}_s + \tilde{B}_sB_r) \tag{7.122}$$

To satisfy equation (7.120) we must have

$$\omega_r \sum_I (\tilde{C}^{rs}_{II} - \tilde{C}^{rs*}_{II}) + \omega_s \sum_I (C^{rs}_{II} - C^{rs*}_{II})$$
$$= \omega_r \sum_{AI} (\tilde{C}^r_{IA}\tilde{C}^s_{AI} - \tilde{C}^r_{AI}\tilde{C}^s_{IA}) + \omega_s \sum_{AI} (C^r_{IA}C^s_{AI} - C^r_{AI}C^s_{IA}) \tag{7.123}$$

Comparing equation (7.123) with the same equation in which we interchanged $r \leftrightarrow s$ we get

$$\sum_I (\tilde{C}_{II}^{rs} - \tilde{C}_{II}^{rs*}) = \frac{\omega_r^2 + \omega_s^2}{\omega_r^2 - \omega_s^2} \sum_{AI} (\tilde{C}_{IA}^r \tilde{C}_{AI}^s - \tilde{C}_{AI}^r \tilde{C}_{IA}^s) + \frac{2\omega_r \omega_s}{\omega_r^2 - \omega_s^2} \sum_{AI} (C_{IA}^r C_{AI}^s - C_{AI}^r C_{IA}^s) \tag{7.124}$$

$$\sum_I (C_{II}^{rs} - C_{II}^{rs*}) = -\frac{2\omega_r \omega_s}{\omega_r^2 - \omega_s^2} \sum_{AI} (\tilde{C}_{IA}^r \tilde{C}_{AI}^s - \tilde{C}_{AI}^r \tilde{C}_{IA}^s) - \frac{\omega_r^2 + \omega_s^2}{\omega_r^2 - \omega_s^2} \sum_{AI} (C_{IA}^r C_{AI}^s - C_{AI}^r C_{IA}^s) \tag{7.125}$$

Equations (7.124) and (7.125) have no meaning for the case $\omega_r = \omega_s$. Therefore, we introduce antisymmetrical relations for the diagonal coefficients

$$C_{PP}^{rs} - C_{PP}^{rs*} = 0 \tag{7.126}$$

and

$$\tilde{C}_{PP}^{rs} - \tilde{C}_{PP}^{rs*} = 0 \tag{7.127}$$

and further we ask

$$\sum_{AI} (C_{IA}^r C_{AI}^s - C_{AI}^r C_{IA}^s) = 0 \tag{7.128}$$

and

$$\sum_{AI} (\tilde{C}_{IA}^r \tilde{C}_{AI}^s - \tilde{C}_{AI}^r \tilde{C}_{IA}^s) = 0 \tag{7.129}$$

Then, according to (7.103) and (7.104), we have

$$C_{PQ}^r = -C_{QP}^r \tag{7.130}$$

$$\tilde{C}_{PQ}^r = \tilde{C}_{QP}^r \tag{7.131}$$

Relations (7.130) and (7.131) are sufficient to satisfy equations (7.128) and (7.129) and therefore to satisfy condition (7.120). Let us further try to make equal to zero the term $H^0_{2(2,\,0)}$; that is, we ask

$$H^0_{2(2,\,0)} = 0 \qquad (7.132)$$

According to (7.75) and (7.82) we get

$$H^0_{2(2,\,0)} = \tfrac{1}{2} \sum_{rs} \left[-V^{rs}_N + \sum_I u^{rs}_{II} + 2 \sum_{AI} (u^r_{IA} + \hbar\omega_r \tilde{C}^r_{IA}) C^s_{AI} \right] B_R B_s \qquad (7.133)$$

$$H^0_{B_{2(2,\,0)}} = 0 \qquad (7.134)$$

This gives us the relation for the new vibrational potential energy V^{rs}_N, which originates from the interaction between the nuclei and the electrons:

$$V^{rs}_N = \sum_I u^{rs}_{II} + \sum_{AI} [(u^r_{IA} + \hbar\omega_r \tilde{C}^r_{IA}) C^s_{AI} + (u^s_{IA} + \hbar\omega_s \tilde{C}^s_{IA}) C^r_{AI}] \qquad (7.135)$$

We can use equation (7.135) to compare the expression for the vibrational potential energies in different representations. In the crude representation for which $C = 0$ and $\tilde{C} = 0$, we get

$$V^{rs}_N = \sum_I u^{rs}_{II} \qquad (7.136)$$

which represents the potential energy of the nuclei in the field of "frozen" electrons. In adiabatic representation for which $\tilde{C} = 0$, we get

$$V^{rs}_N = \sum_I u^{rs}_{II} + \sum_{AI} (u^r_{IA} C^s_{AI} + u^s_{IA} C^r_{AI}) \qquad (7.137)$$

which represents the potential energy of nuclei in the field of electrons that adiabatically follow the motion of nuclei. In the case when the conditions for the validity of adiabatic approximation are not satisfied, we have to use equation (7.135) which represents the potential energy of nuclei in the field of electrons that do not follow the motion of nuclei and that move with the phase shift. As a last step we make equal to zero the terms $H^0_{2(0,\,2)}$, that is,

$$H^0_{2(0,\,2)} = 0 \qquad (7.138)$$

According to (7.76) and (7.83) we have

$$H^0_{A2(0,2)} = \sum_{rs}\left(-\tfrac{1}{2}W^{rs}_N - \sum_{AI}\hbar\omega_r C^r_{AI}\tilde{C}^s_{IA}\right)\tilde{B}_r\tilde{B}_s \tag{7.139}$$

and

$$H^0_{B2(0,2)} = 2\sum_{AIrs}\hbar\omega_r C^r_{AI}\tilde{C}^s_{IA}\tilde{B}_r\tilde{B}_s \tag{7.140}$$

which gives us the kinetic energy W^{rs}_N

$$W^{rs}_N = 2\hbar\omega_r\sum_{AI}C^r_{AI}\tilde{C}^s_{IA} \tag{7.141}$$

This kinetic energy term has no analogy in either the adiabatic representation or in the crude representation. This term originates from the electrons which do not follow adiabatically the motion of the nuclei. Together with the nuclei, the kinetic energy term forms the total vibrational kinetic energy. Knowing the potential energy V^{rs}_N (7.135) and kinetic energy W^{rs}_N (7.141) we can substitute them into the vibrational Hamiltonian H_B in (7.36), and we can try to solve the vibrational equations. Note that both these quantities depend on the frequencies ω_r; therefore, we try to suggest the iterative procedure to solve the vibrational equations in nonadiabatic representation.[36] Let us write the vibrational Hamiltonian H_B in (7.31)

$$H_B = \sum_r \hbar\omega_r(\bar{b}^+_r\bar{b}_r + 1/2) = \frac{\hbar}{4}\sum_r \omega_r(\bar{B}_r\bar{B}_r - \bar{\bar{B}}_r\bar{\bar{B}}_r) \tag{7.142}$$

as a sum of the kinetic energy part and the potential energy part

$$H_B = \tfrac{1}{2}\sum_{rs}E^{rs}_{\text{kin}}\bar{\bar{B}}_r\bar{\bar{B}}_s + \tfrac{1}{2}\sum_{rs}E^{rs}_{\text{pot}}\bar{B}_r\bar{B}_s \tag{7.143}$$

According to the equipartition for the harmonic oscillator we have

$$E^{rs}_{\text{kin}} = T^{rs}_N + W^{rs}_N = -\tfrac{1}{2}\hbar\omega_r\,\delta_{rs} \tag{7.144}$$

$$E^{rs}_{\text{pot}} = E^{rs}_{NN} + V^{rs}_N = \tfrac{1}{2}\hbar\omega_r\,\delta_{rs} \tag{7.145}$$

Let us now express the kinetic and potential energies in some different basis set of coordinate and momentum operators as

$$E_{\text{kin}} = \tfrac{1}{2} \sum_{rs} \overset{0}{E}{}^{rs}_{\text{kin}} \overset{0}{\bar{\bar{B}}} \overset{0}{\bar{\bar{B}}}_s \tag{7.146}$$

$$E_{\text{pot}} = \tfrac{1}{2} \sum_{rs} \overset{0}{E}{}^{rs}_{\text{pot}} \overset{0}{\bar{B}}_r \overset{0}{\bar{B}}_s \tag{7.147}$$

where generally for the matrix elements $\overset{0}{E}{}^{rs}_{\text{kin}}$ and $\overset{0}{E}{}^{rs}_{\text{pot}}$ relations (7.144) and (7.145) need not to be satisfied. Both basis sets, if they are complete, are given by the transformation

$$\overset{0}{\bar{B}}_r = \sum_s \alpha_{rs} \bar{B}_s \tag{7.148}$$

$$\overset{0}{\bar{\bar{B}}}_r = \sum_s \beta_{rs} \bar{\bar{B}}_s \tag{7.149}$$

The transformation matrices must satisfy the following symmetry relations

$$\alpha_{rs} = \alpha^*_{rs} \tag{7.150}$$

and

$$\beta_{rs} = \beta^*_{rs} \tag{7.151}$$

The commutation relations between coordinate and momentum operators give us

$$[\bar{B}_r, \bar{\bar{B}}_s] = [\overset{0}{\bar{B}}_r, \overset{0}{\bar{\bar{B}}}_s] = -2\delta_{rs} \tag{7.152}$$

This brings us to the following condition for the matrices of transformation α and β:

$$\sum_t \alpha_{rt} \beta^*_{st} = \delta_{rs} \tag{7.153}$$

Substituting (7.148) and (7.149) into (7.146) and (7.147) and comparing with (7.143)–(7.145), we get

$$\frac{2}{\hbar}\sum_{rs} \overset{0}{E}{}^{rs}_{\text{kin}}\,\beta_{rt}\beta_{su} = -\omega_t\,\delta_{tu} \tag{7.154}$$

and

$$\frac{2}{\hbar}\sum_{rs} \overset{0}{E}{}^{rs}_{\text{pot}}\,\alpha_{rt}\,\alpha_{su} = \omega_t\,\delta_{tu} \tag{7.155}$$

From equations (7.154) and (7.155) we get

$$\frac{4}{\hbar^2}\sum_{rst} \overset{0}{E}{}^{rs}_{\text{kin}}\,\overset{0}{E}{}^{st}_{\text{pot}}\,\beta_{rv}\,\alpha_{tu} = -\omega_u^2\,\delta_{vu} \tag{7.156}$$

from which, after simple manipulation, we get the characteristic equation for the vibrational frequencies:

$$\sum_t \left(\frac{4}{\hbar^2}\sum_s \overset{0}{E}{}^{rs}_{\text{kin}}\,\overset{0}{E}{}^{st}_{\text{pot}} + \omega_u^2\,\delta_{rt}\right)\alpha_{tu} = 0 \tag{7.157}$$

The normalization condition for the vectors α we obtain from equation (7.155)

$$\sum_{rs} \overset{0}{E}{}^{rs}_{\text{pot}}\,\alpha^*_{rt}\,\alpha_{st} = \frac{\hbar}{2}\,\omega_t \tag{7.158}$$

The iterative procedure to solve equation (7.157) can be suggested as follows. To the matrix elements of energies $\overset{0}{E}_{\text{kin}}$ and $\overset{0}{E}_{\text{pot}}$ we associate values $E_{\text{kin}(k-1)}$ and $E_{\text{pot}(k-1)}$. Then the kth iteration step will be done by equation (7.157) from which we get the vibrational frequencies $\omega_{(k)}$ and vectors $\alpha_{(k)}$, which we normalize according to (7.158). From the values $\alpha_{(k)}$ using equation (7.153) we get $\beta_{(k)}$. Matrix $\alpha_{(k)}$ permits us to transform the potential energy $E_{NN(k-1)}$ onto $E_{NN(k)}$ and the matrix elements of electron–phonon interaction $u_{(k-1)}$ onto $u_{(k)}$. Using the matrix $\beta_{(k)}$ we can transform the kinetic energy $T_{N(k-1)}$ onto $T_{N(k)}$. Then using equations (7.103) and (7.104) we get the new coefficients $C_{(k)}$ and $\tilde{C}_{(k)}$. Finally we calculate $V_{N(k)}$ and

$W_{N(k)}$ using (7.135) and (7.141) and then $E_{\text{kin}(k)}$ and $E_{\text{pot}(k)}$. If the iteration procedure converges

$$\lim_{k \to \infty} \alpha_{rs(k)} = \delta_{rs} \tag{7.159}$$

and

$$\lim_{k \to \infty} \omega_{rs(k)} = \omega_{rs} \tag{7.160}$$

Let us comment on the center-of-mass problem. Very important for the described iteration procedure is the first iteration step. Here it comes to the transformation from nuclei coordinates to normal coordinates and therefore to the separation of the motion onto the vibrational, translational, and rotational. This separation is fundamental for the center-of-mass problem. Note that equations (7.144), (7.145), (7.135), and (7.141) are fully vibrational equations, and they respect the center-of-mass problem. These equations cannot be solved explicitly but only iteratively, and therefore the first iteration step is fundamental. It would be simple to pass from nuclear coordinates to normal coordinates in the first iteration step on the adiabatic level, but in this way we arrive at the adiabatic center of mass. Therefore, we leave open here the problem of how to make the first iteration step.

To make the equations for the C and $\tilde{C}$ coefficients (7.103) and (7.104) as well as the equation for the equilibrium position of the nuclei (7.119) and the equations for the vibrational potential energy (7.135) and vibrational kinetic energy (7.141) more practical for the numerical calculations, we simplify these equations by the use of *CPHF* equations.[16] When discussing the connection of the technique of canonical transformations with the analytical derivative methods, we introduced the following basis set from (5.21)

$$|\hat{P}\rangle = |\hat{P}(R)\rangle = \sum_{\mu} \gamma_{P_\mu}(0)|\mu(R)\rangle \tag{7.161}$$

which is practical for the solution of the *CPHF* equations. We can also use this basis set to simplify the above-mentioned equations in nonadiabatic representation. Let us write them again. The equations for C and $\tilde{C}$ coefficients (7.103) and (7.104) can be simplified using the known symmetry relations between C^r_{PQ} and $\tilde{C}^r_{PQ}$ coefficients as follows

$$u^r_{PQ} + (\epsilon^0_P - \epsilon^0_Q)C^r_{PQ} + \sum_{AI} (2v^0_{PAQI} - v^0_{PAIQ} - v^0_{PIAQ})C^r_{AI} - \hbar\omega_r \tilde{C}^r_{PQ} = \epsilon^r_P\, \delta_{PQ} \tag{7.162}$$

and

$$(\epsilon_P^0 - \epsilon_Q^0)\tilde{C}_{PQ}^r + \sum_{AI} (v_{PAIQ}^0 - v_{PIAQ}^0)\tilde{C}_{AI}^r - \hbar\omega_r C_{PQ}^r = 0 \qquad (7.163)$$

The equations for the equilibrium position of nuclei (7.119), for the potential (7.135), and for the kinetic vibrational energy (7.141) are, respectively,

$$E^r = E_{NN}^r + \sum_I u_{II}^r = 0 \qquad (7.164)$$

$$V_N^{rs} = \sum_I u_{II}^{rs} + \sum_{AI} [(u_{AI}^r + \hbar\omega_r \tilde{C}_{AI}^r)C_{AI}^s + (u_{AI}^s + \hbar\omega_s \tilde{C}_{AI}^s)C_{AI}^r] \qquad (7.165)$$

$$W_N^{rs} = 2\hbar\omega_r \sum_{AI} C_{AI}^r \tilde{C}_{AI}^s \qquad (7.166)$$

Let us write the first derivative of the adiabatic Hartree–Fock operator in (4.13) with respect to the normal coordinates. We have

$$f_{PQ}^r = u_{PQ}^r + (\epsilon_P^0 - \epsilon_Q^0)C_{PQ}^r + \sum_{AI} (2v_{PAQI}^0 - v_{PAIQ}^0 - v_{PIAQ}^0)C_{AI}^r \qquad (7.167)$$

Using equation (7.167) we write equation (7.162) in the following way:

$$f_{PQ}^r - \hbar\omega_r \tilde{C}_{PQ}^r = \epsilon_P^r \, \delta_{PQ} \qquad (7.168)$$

Using the basis set (7.161) we have the operator f_{PQ} given also by equation (5.28). Calculating the first derivative of (5.28) we get

$$f_{PQ}^r = \hat{f}_{PQ}^r + (\epsilon_P^0 - \epsilon_0^0)\hat{C}_{PQ}^r + \sum_{RI} (2v_{PRQI}^0 - v_{PRIQ}^0 - v_{PIRQ}^0)\tilde{C}_{RI}^r - \epsilon_Q^0 \hat{S}_{PQ}^r \qquad (7.169)$$

Substituting (7.169) into (7.168) we get

$$\hat{f}_{PQ}^r + (\epsilon_P^0 - \epsilon_0^0)\hat{C}_{PQ}^r + \sum_{RI} (2v_{PRQI}^0 - v_{PRIQ}^0 - v_{PIRQ}^0)\hat{C}_{RI}^r - \epsilon_Q^0 \hat{S}_{PQ}^r - \hbar\omega_r \tilde{C}_{PQ}^r = \epsilon_P^r \, \delta_{PQ} \qquad (7.170)$$

Substituting (5.24) into (7.163) we obtain

$$(\epsilon_P^0 - \epsilon_Q^0)\tilde{C}_{PQ}^r + \sum_{AI} (v_{PAIQ}^0 - v_{PIAQ}^0)\tilde{C}_{AI}^r - \hbar\omega_r(\hat{C}_{PQ}^r + \langle P|\hat{Q}^r\rangle) = 0 \qquad (7.171)$$

In a similar way we can simplify equations (7.164)–(7.166). Calculating the first derivative of the E_{SCF} in (4.12) with respect to the normal coordinates we have

$$E^r_{SCF} = \sum_I u^r_{II} \tag{7.172}$$

Substituting (7.172) into (7.164) we get

$$E^r = E^r_{NN} + E^r_{SCF} = 0 \tag{7.173}$$

Calculating the first derivative of E_{SCF} (5.27) we get

$$E^r_{SCF} = \hat{E}^r_{SCF} - \sum_I \epsilon^0_I \hat{S}^r_{II} \tag{7.174}$$

Substituting (7.174) into (7.173) we obtain

$$E^r = E^r_{NN} + \hat{E}^r_{SCF} - \sum_I \epsilon^0_I \hat{S}^r_{II} = 0 \tag{7.175}$$

Calculating the second derivative of adiabatic E_{SCF} in (4.12) with respect to the normal coordinates and using (7.162) we get

$$E^{rs}_{SCF} = \sum_I u^{rs}_{II} + \sum_{AI} [(u^r_{AI} + \hbar\omega_r \tilde{C}^r_{AI}) C^s_{AI} + (u^s_{AI} + \hbar\omega_s \tilde{C}^s_{AI}) C^r_{AI}] \tag{7.176}$$

Then we can write the expression for V^{rs}_N (7.165) in this simple way:

$$V^{rs}_N = E^{rs}_{SCF} \tag{7.177}$$

Calculating the second derivative E^{rs}_{SCF} from equation (5.27) and using (7.170) we finally get for the vibrational potential energy the expression

$$\begin{aligned} V^{rs}_N = \hat{E}^{rs}_{SCF} &- \sum_I [\epsilon^0_I \hat{S}^{rs}_{II} + \tfrac{1}{2}(\epsilon^r_I \hat{S}^s_{II} + \epsilon^s_I \hat{S}^r_{II})] \\ &+ \sum_{RI} [(\hat{f}^r_{RI} - \epsilon^0_I \hat{S}^r_{RI} + \hbar\omega_r \tilde{C}^r_{RI}) \hat{C}^s_{RI} + (\hat{f}^s_{RI} - \epsilon^0_I \hat{S}^s_{RI} + \hbar\omega_s \tilde{C}^s_{RI}) \hat{C}^r_{RI}] \end{aligned} \tag{7.178}$$

For the vibrational kinetic energy in (7.166), substituting from (5.24) we get

$$W_N^{rs} = 2\hbar\omega_r \sum_{AI} (\hat{C}^r_{AI} + \langle A|I^r\rangle)\hat{C}^s_{AI} \tag{7.179}$$

We can look at equations (7.170), (7.175), (7.178), and (7.179) as the generalization of the *CPHF* equations for the case of nonadiabatic representation.

In the adiabatic approximation in which $\tilde{C} = 0$, we get from (7.170) and from (7.178) the very well known *CPHF* equations[16]

$$\begin{aligned} \hat{f}^r_{PQ} &+ (\epsilon^0_P - \epsilon^0_Q)\hat{C}^r_{PQ} \\ &+ \sum_{RI} (2\overset{0}{v}_{PRQI} - \overset{0}{v}_{PRIQ} - \overset{0}{v}_{PIRQ})\hat{C}^r_{RI} - \epsilon^0_Q \hat{S}^r_{PQ} = \epsilon^r_P\, \delta_{PQ} \end{aligned} \tag{7.180}$$

and

$$\begin{aligned} \overset{0}{V}{}^{rs}_N = \hat{E}^{rs}_{SCF} &- \sum_I [\epsilon^0_I \hat{S}^{rs}_{II} + \tfrac{1}{2}(\epsilon^r_I \hat{S}^s_{II} + \epsilon^s_I \hat{S}^r_{II})] \\ &+ \sum_{RI} [(\hat{f}^r_{RI} - \epsilon^0_I \hat{S}^r_{RI})\hat{C}^s_{RI} + (\hat{f}^s_{RI} - \epsilon^0_I \hat{S}^s_{RI})\hat{C}^r_{RI}] \end{aligned} \tag{7.181}$$

References

1. P. Pulay, in: *Ab Initio Methods in Quantum Chemistry, Vol. II* (K. P. Lawley, ed.), Wiley (1987).
2. E. A. Salter, G. W. Trucks, and R. J. Bartlett, *J. Chem. Phys.* **90**, 1752 (1989).
3. J. A. Pople, in: *Geometrical Derivatives of Energy Surfaces and Molecular Properties* (P. Jorgensen and J. Simons, eds.), p. 109, Reidel, Dordrecht (1986).
4. N. H. March, W. H. Young, and S. Sampanthar, *The Many-Body Problem in Quantum Mechanics*, Cambridge University Press (1967).
5. J. Čížek, *J. Chem. Phys.* **45**, 4256 (1966).
6. J. Čížek, *Adv. Chem. Phys.* **14**, 35 (1969).
7. J. Paldus and J. Čížek, *Adv. Quant. Chem.* **9**, 105 (1975).
8. J. A. Pople, J. S. Binkley, and R. Seeger, *Int. J. Quant. Chem. Symp.* **10**, 1 (1976).
9. I. Hubač and P. Čársky, *Topics Curr. Chem.* **75**, 97 (1978).
10. R. J. Bartlett, *Annu. Rev. Phys. Chem.* **32**, 359 (1981).
11. S. Wilson, *Electron Correlation in Molecules*, Clarendon Press, Oxford (1984).
12. K. Jankowski, in: *Methods in Computational Chemistry, Vol. 1: Electron Correlation in Atoms and Molecules* (S. Wilson, ed.), p. 1, Plenum, New York (1987).
13. M. Urban, J. Černušák, V. Kellö, and J. Noga, in: *Methods in Computational Chemistry, Vol. 1: Electron Correlation in Atoms and Molecules* (S. Wilson, ed.), p. 117, Plenum, New York (1987).
14. J. Gerratt and J. M. Mills, *J. Chem. Phys.* **49**, 1719 (1968).
15. J. Gerratt and J. M. Mills, *J. Chem. Phys.* **49**, 1730 (1968).

16. J. A. Pople, K. Raghavachari, H. B. Schlegel, and J. S. Binkley, *Int. J. Quantum Chem. Symp.* **13**, 225 (1979).
17. N. C. Handy, R. D. Amos, J. F. Graw, J. E. Rice, and E. D. Simandiras, *Chem. Phys. Lett.* **120**, 151 (1985).
18. R. J. Bartlett, in: *Geometrical Derivatives of Energy Surfaces and Molecular Properties* (P. Jorgensen and J. Simons, eds.), p. 35, Reidel, Dordrecht (1986).
19. E. A. Salter and R. J. Bartlett, *J. Chem. Phys.* **90**, 1967 (1989).
20. P. Jorgensen and T. Helgaker, *J. Chem. Phys.* **89**, 1560 (1988).
21. T. Helgaker, P. Jorgensen, and N. C. Handy, *Theor. Chim. Acta* **76**, 227 (1989).
22. C. Moller and M. S. Plesset, *Phys. Rev.* **46**, 618 (1934).
23. I. Hubač and M. Svrček, *Int. J. Quant. Chem.* **23**, 403 (1988).
24. I.Hubač, M. Svrček, E. A. Salter, C. Sosa, and R. J. Bartlett, *Lecture Notes in Chemistry*, Vol. 52, p. 95, Springer, Berlin (1988).
25. J. P. Blaizot and G. Ripka, *Quantum Theory of Finite Systems*, MIT Press, Cambridge, MA (1986).
26. M. Hamermesh, *Group Theory and its Application to Physical Problems*, Addison-Wesley, Reading, MA (1964).
27. I. Hubač, V. Kvasnička, and A. Holubec, *Chem. Phys. Lett.* **23**, 381 (1973).
28. I. Hubač and M. Urban, *Theoret. Chim. Acta* **45**, 185 (1977).
29. M. Svrček and I. Hubač, *Czech. J. Phys.* **41**, 556 (1991).
30. R. J. Harrison, G. B. Fitzgerald, W. D. Laiding, and R. J. Bartlett, *Chem. Phys. Lett.* **124**, 291 (1986).
31. B. H. Brandow, *Rev. Mod. Phys.* **39**, 771 (1967).
32. B. H. Brandow, *Int. J. Quant. Chem.* **15**, 207 (1979).
33. J. Lindgren, *Int. J. Quant. Chem. Symp.* **12**, 33 (1978).
34. V. Kvasnička, *Adv. Chem. Phys.* **36**, 345 (1977).
35. M. Svrček and I. Hubač, *Int. J. Quant. Chem.* **31**, 625 (1987).
36. M.Svrček, Ph.D. Thesis, Faculty of Mathematics and Physics, 1986, Bratislava, Czechoslovakia.
37. T. H. Dunning, Jr., *J. Chem. Phys.* **53**, 2823 (1970).
38. L. T. Redmon, G. D. Purvis III, and R. J. Bartlett, *J. Am. Chem. Soc.* **101**, 2856 (1979).
39. E. Davidson and D. Feller, *Chem. Phys. Lett.* **104**, 54 (1984).
40. W. Kolos and L. Wolniewicz, *J. Chem. Phys.* **49**, 409 (1988).
41. L. Wolniewicz, *J. Chem. Phys.* **78**, 6173 (1983).
42. M. Born and R. Oppenheimer, *Annl. Phys.* (*Leipzig*) **84**, 4357 (1927).
43. H. Köppel, W. Domcke, and L. S. Cederbaum, *Adv. Chem. Phys.* **57**, 59 (1984).
44. M. Wagner, *Phys. Stat. Sol.*(*b*) **107**, 617 (1981).

Contents of Previous Volumes

Volume 3: Concurrent Computation in Chemical Calculations

Author Index

Subject Index